Robot Technology

Volume 3B: Teleoperation and Robotics

Robot Technology

A Series in Eight Volumes

Series Editor and Consultant: Philippe Coiffet
English Language Series Consultant:
I. Aleksander, Imperial College of Science and Technology, London, England

Volume 3B

TELEOPERATION AND ROBOTICS

Applications and Technology

Jean Vertut and Philippe Coiffet

HERMES PUBLISHING
London – Paris – Lausanne

This work entitled *Teleoperation and Robotics* is published as two volumes:

Volume 3A *Evolution and Development*
Part 1 Introduction

Volume 3B *Applications and Technology*
Part 2 The contribution of computer science
Part 3 Performance and the man-machine interface
Part 4 Applications of teleoperation

The references cited in Volume 3A are placed after the text in Volume 3A, and the complete bibliography is included in Volume 3B. Both volumes are indexed separately. Whenever possible the figures are numbered consecutively according to the section in which they appear.

Translated by Meg Tombs

First published 1985 by Hermes Publishing (France)
51 Rue Rennequin, 75017 Paris, France

English edition first published 1985
by Kogan Page Ltd, 120 Pentonville Road, London N1 9JN

ISBN 1 85121 003 2

Printed and bound in Great Britain by
Robert Hartnoll (1985) ltd., Bodmin, Cornwall

Contents

Publisher's note

It was shortly before going to press that we learnt of the death of Jean Vertut, the author of *Teleoperation and Robotics*. It is our hope that, in publishing these volumes, we have helped to provide a fitting tribute to his outstanding contribution to the field of teleoperation.

Foreword

It is a privilege to be asked to introduce this important work. Such a book has long been needed. Industrial manipulators and robots have caught the attention of the general public and become very fashionable in the last few years. The casual reader of current newspapers and magazines or the viewer of television and films might easily conclude that the development of mechanical hands, arms and legs or other mobility devices has progressed rapidly in only the last few years. Most people are unaware of the gradual orderly succession of creative designs and painstaking refinements which have been produced over a greater number of years.

That story is carefully described in this volume, together with diagrams and photographs which document in detail this elegant phase in the history of machine design. This volume together with Volume 3A constitute the most complete and comprehensive work on manipulators and teleoperators. Jean Vertut and Philippe Coiffet are well known not only as authors but also as engineers who have produced some of the finest devices in the world.

Of course for the complete history of manipulators and teleoperators one must look back to the artisans who crafted the delightful clockworks, mechanical puppets and toys before and during the Renaissance. However, the modern development of these devices was spurred most by the need which arose, concurrent with the development of nuclear power, to handle radioactive materials remotely. Unquestionably, 30 years ago the group at Argonne National Laboratory near Chicago headed by Ray Goertz led the field in mechanical design of manipulators. Shortly afterwards, in my judgement, world leadership swung to the CEA laboratory of Jean Vertut and during recent years the French School has set the pace – with intense competition from the Japanese and perhaps more recently the Americans.

The engineer or industrialist interested in automatic and computer-based machinery which is general purpose, ie which can be programmed to perform a great variety of tasks, should realize the close link with the design of machines to aid and extend the human hand for remote manipulation. This is not to ignore the mechanical design art in fixed automation which has accumulated since the industrial revolution.

It is only to appreciate the degree to which biomechanics and artificial mechanics interact to enrich understanding and catalyse new designs.

As we look ahead to designing manipulators, teleoperators and robots which integrate artificial sensors and intelligence with mechanical hands, arms and legs, it is important to keep in mind the lessons of man-machine symbiosis so elegantly presented in this work.

Thomas B. Sheridan
Professor of Engineering and Applied Psychology
Department of Mechanical Engineering
Massachusetts Institute of Technology
January 1985

Preface

This work is the result of 10 years of friendly and far-reaching co-operation between Philippe Coiffet and myself. This period represents a good third of the time I have devoted to remote handling and, more recently, robotics.

In addition to having been a member of the nuclear research community for some time and the influence of Ray Goertz, whom I met many times between 1962 and 1969, and his team on my work, I am also privileged to be involved in the study of work in all types of hostile environment, such as under water and outer space, aids to the handicapped, civil protection and industrial robotics. I share this privilege with Carl Flatau, who developed the use of teleoperation in particle accelerators, touched upon the use of teleoperation in outer space and in industrial robotics, and with whom I collaborated briefly at the time when our first electronic manipulator was developed. This study was productive for both parties.

A collaboration with Hans Kleinwachter, who made the transition from working in the field of outer space manipulation to working in the field of nuclear teleoperation during the Syntelman project, which remained at experimental level, was also productive. The innovators in the field of teleoperation are very small in number and include Ralph Mosher, who chaired the session when I delivered my first paper in the USA in 1962.

This liaison has led to my being involved in two scientific communities, both of which are featured largely in the bibliography of this work. The first body is the Remote System Technology Division (RSTD), which today forms part of the American Nuclear Society (ANS), and was originally established well in advance of the latter group by Ray Goertz in 1952. I have belonged to this community since 1962. The second body is the International Federation for Theory of Machines and Mechanisms (IFToMM), which France joined as late as 1979 in connection with its work on robotics. Since 1973 the IFToMM has organized, with the Centre International des Sciences Mécaniques (CISM), Udine, Italy, Robot and Manipulator Symposiums (RO.MAN.SY). These symposiums provide the only venue for world-wide exchange between roboticists, including mechanical engineers, and those delegates from Eastern countries. *RO.MAN.SY Proceedings of the Fifth Symposium on Theory and Practice of Robots and Manipulators* Udine, Italy, June, 1984, edited by A. Morecki, G. Bianchi and K. Kedzior is available from Kogan Page.

Edwin G. Johnsen is another pioneer extensively quoted in the bibliography of this work, who, at the time of the American nuclear-powered rocket project NERVA, directed a team made up of representatives from NASA and the Atomic Energy Commission (AEC). He introduced the term *teleoperator* to mean a mobile telemanipulator which is not directly mechanically linked with the operator, whereas other terms such as *telechirics* (coined by M.W. Thring, Great Britain), *telemation* (used by Alexander, NASA, USA) and *telesymbiotics* (devised by Joel Charles, France) have not been so widely adopted.

Another new term, *telepresence*, is intended to describe sensory feedback, while the word *telesymbiotics* refers to the human operator, who is present (via the teleoperator) in the remote environment, and has these sensory, physical and decision making abilities. According to the authors, however, telepresence can function either unilaterally or bilaterally, and is therefore equivalent to telesymbiotics.

The importance of Edwin Johnsen's work remains undiminished, even after 15 years, which is why our reference to works published before 1967 are the ones he did not cite himself.

Over the last 10 years, I have also collaborated well with Joel Charles and the team at the Technical Direction for Naval Construction (DTCN), who have worked on developing the technology for work at depths of 6,000 m, built the first French Bathyscaphe, Archimede, and are currently involved with the CNEXO (now called IFREMER) SM97 submarine project. This was so named because 97 per cent of the sea bed is accessible at 6,000 m submergence.

My good friend Melvin Feldman is in charge of the nuclear telemaintenance team at Oak Ridge. This opens up new prospects for Franco-American exchange on advanced teleoperation. As a result of this, we both tried offering, in recent years, what help we could to colleagues at Three Mile Island, for whom teleoperation should provide a tremendous help in the recovery of their TMI-2 nuclear reactor, which has made slow progress since the 1979 accident.

As a result of a fascinating encounter with P. Rabischong, who introduced me to the field of biomechanics (which he defines as 'scientific espionage of living mechanical systems'), I discovered the Laboratory for Automation and Microprocessing, Montpellier (LAMM). This prompted the collaboration with Philippe Coiffet on the technology of (tele)manipulators and research into their control. This programme between an academic laboratory and a team from a large technical organization started over 10 years ago, and at the same time the friendship between my co-author and myself was founded, resulting today in our awareness of and activity in the new field of service robotics.

At the time Edwin Johnsen's first works were appearing, our friend and colleague Michel Grenon published *Le travail en milieu hostile* (*Work in hostile environments*), which until now was the only book written in French on the subject (Grenon, 1968). More recently, our colleague W. Köhler (West Germany) produced an immense work containing all available information on existing telemanipulators, and his work is also frequently mentioned in the bibliography.

This work was devised by Philippe Coiffet and myself. Coiffet handled Part 2 on the contribution of computer science, a new aspect in teleoperation, as well as Chapter 3.1 on performance evaluation of teleoperation systems. Nicole Fiori and James Richardson of the Physiology of Movement Laboratory, Orsay Faculty, wrote Chapter 3.2 in Part 3, which concerns the human operator in relation to teleoperation systems. I wrote Parts 1 and 4. Part 1 is the entire text for Volume 3A and gives an introduction and account of the development of teleoperation. Part 4 describes a number of applications.

At a time when multidisciplinary analysis is being again adopted, after centuries of specialization, we have attempted to present a broad approach to the subject which, to a greater extent than many others, refuses to be restricted by an over-specialized approach. Therefore, in Part 4 I have summarized in terms of the different applications, the semi-historical progress of the preceding parts in terms of function, and suggested links between different areas of application. Although I may have succumbed to the temptation of writing at length on the subject of nuclear applications, this was in the hope of making them more widely known and helping to demystify this area as a whole, particularly in dispelling the impression that it is impossible to tackle dangerous situations involving nuclear materials by hand, which is entirely possible using remote handling techniques. Part 4 also cites the most recent work in the field.

I should like to thank all those without whom this work would not have been possible, the pioneers first of all, and in particular those who never had the opportunity to witness the possibilities opened up by advanced teleoperation, that is the advance from telemanipulation to teleoperation, and then to the wider context of robotics: my teachers and in particular André Leroi-Gourhan and Marcel Jousse (1965) who, in the fields of prehistory and anthropology taught me the importance of tackling problems directly and with a global approach (which is so difficult); next in chronological order, my closest collaborators J.-C. Germond, J. Le Tohic, R. Séran, J.-P. Guilbaud, P. Marchal, my close colleagues and friends P. Pesanti, G. Lefort, P. Auchapt, A. Crégut, J. Guittet, and more recently J.-L. Rouyer; outside the CEA P. Rabischong, J. Charles, P. Coiffet, A. Liégeois, A. Pardo, B. Espiau, S. Bouisset, J.-C. Guinot; in alphabetical order, in Germany,

H. Kleinwachter and W. Köhler; in the USA A. Bejczy, C. Flatau, J. Hill, B. McGhee, B. Roth, K. Salisbury, T. Sheridan, particularly M. Feldman, and the late J. Simon; in Japan, I. Kato; in Poland, A. Morecki; in the USSR, the late Artobolevski, and A. Bessonov; and finally, those working in industry producing products designed by the researchers: the late J.-C. Mettetal and C. Piron, in the USA; F. Chesley, D. Jelatis and particularly J.-P. Cazalis, whose team has been responsible for the production and distribution of much of my work.

It is not possible to mention all the members of the young team who are now working, under the direction of J.-L. Rouyer, P. Marchal and G. Fraize, with their friends in other teams on the subjects mentioned in this volume.

Finally, I should like to thank all those, at the CEA and elsewhere, who have contributed so much: P. Balligand, B. Jacquard, P. Tanguy, J. Pradel, and recently A. Milliés.

Jean Vertut
January 1985

The object behind Volumes 3A and 3B in the *Robot Technology* series is to show two aspects of a discipline called teleoperation, which until recently was not included in the field of robotics. It concerns the exploration and exploitation of spaces which because of their inaccessibility or hostility are generally forbidden to man. The first aspect covers the advances made in mastering teleoperation, using only a knowledge of mechanics. The second aspect concerns the more recent contribution of computer science and automatic control.

It was not until 1972, having written a thesis on another field of interest, particle accelerators, that I founded a research team into robotics at Montpellier. This same year I was introduced to teleoperation and met Jean Vertut, its driving force.

What struck me at the time was the extraordinary lack of communication between the world of industrial robotics for manufacturing systems (where automatic control is a basic necessity) and the world of nuclear teleoperation (in which man controls the system from a control station and mechanical solutions are investigated).

Working myself in the field of automatic control, the close collaboration with Jean Vertut, whose encyclopaedic knowledge of teleoperation is truly astonishing, was evidently very productive. This was enhanced by our solid friendship, and our common work well proved this point.

The problem with encyclopaedic knowledge is to turn it into a written account that reflects the genius behind it. This explains the rather lengthy process of completing the text, which started as a project of 200 to 300 pages, and has ended up as a two volume work running to more than twice the original length. However, this study is exhaustive, particularly concerning the history of the field.

The reader should be aware that in the last few years the approach adopted in industrial robotics (in which a simple task is automated first, then made more complex as the technology advances) and the approach used in teleoperation (in which complex tasks must be carried out, and technical advances have increased the efficiency with which the tasks are performed) tend to merge, so that in about 20 years' time, using in addition artificial intelligence techniques, near-perfect robots can be manufactured.

I should like to thank all the people with whom I have worked in this field, both at Montpellier and elsewhere, and in particular A. Liégeois, A. Fournier, B. Espiau, P. Marchal, M. Chirouze, E. Dombre, M.J. Aldon, R. Fournier, P. Rives, J. Pot, P. Kuspriyanto, R. Zapata, P. Dauchez, B. Jouvencel and S. Tobarghan.

Philippe Coiffet
January 1985

Part 2
The contribution of computer science

Chapter 2.1

A description of teleoperation systems

2.1.1 Introduction

Teleoperation used in conjunction with computer science can improve performance. However, in order for the system to operate efficiently, the computer must store models of the system it is intended to control. This is one of the principles of all types of automation using a computer.

In this chapter, the parts that constitute a teleoperation system, and their interrelationships, are described. It is easy to appreciate the functional complexity of the systems considered, and hence how difficult they are to master. Therefore, *teleoperation with mechanical transmission, teleoperation with electrical transmission* and *computerized teleoperation* are described, and diagrams are given for each system.

2.1.2 Teleoperation with mechanical transmission

As has already been described (see Volume 3A), this system usually involves remote manipulation through a shielding (as the movements are transmitted mechanically), which allows the operator to view the slave environment through a transparent window.

2.1.2.1 THE PARTS [see Figure 2.1.2.(1)]

The parts of the teleoperation system are spread over three environments:

1. The *master environment* which houses the human operator and the control system (known as the arm or master system). In mechanical remote manipulation, the latter will, of necessity and for obvious reasons, be an articulated mechanical structure with the same number of degrees of freedom (DOF) as the structure under remote control (the slave arm), and with the same geometry [see Chapter 1.3 and Figure 1.3.0.(2)].
2. The *slave environment* which corresponds to the work area situated on the other side of the shielding where, in particular, the slave arm or slave system is found. To avoid confusion, the term

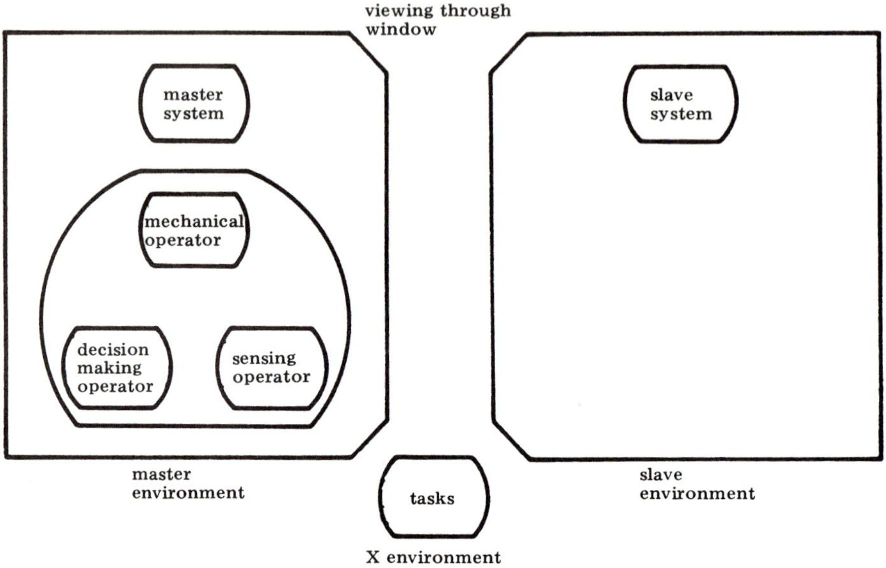

Figure 2.1.2.(1) *The three parts that comprise the mechanical teleoperation system*

master system will be used to describe the master side of the teleoperation system and *master environment* to describe the area in which it works. Similarly, *slave system* and *slave environment* define the teleoperation system and the work area on the slave side respectively.

3. Non-fixed and not necessarily spatial environment, which can be termed the *X environment.* It will contain elements necessary to the understanding of the task description. These elements may not be physical (software), and may be situated in the slave environment, the master environment or elsewhere (in the computer in computerized teleoperation systems). In mechanical remote manipulation, the X environment contains a type of description of the tasks to be accomplished in the slave environment. In reality, this description is situated in the human operator's mind.

To help understand the system, the roles of the human operator can be divided into three groups: mechanical actions, sensing abilities and decision making abilities. The properties of each category are described below [see Figure 2.1.2.(2)].

1. The *mechanical operator* is simply an insensitive generator of physical actions, carrying out the commands received solely from

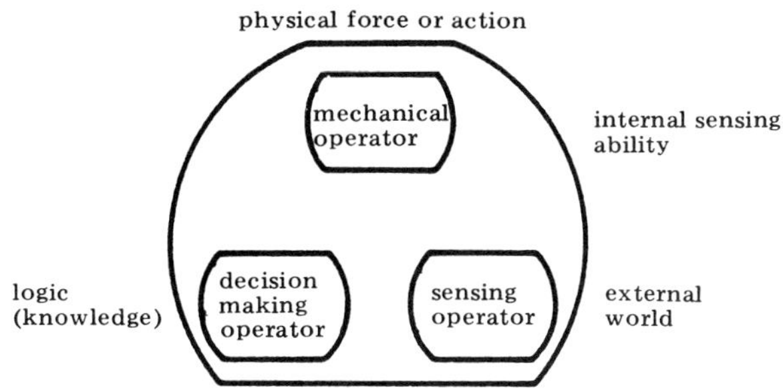

Figure 2.1.2.(2) *The functions of the human operator can be divided into three groups*

the decision making operator (the brain). These commands all involve a force which is converted (or not as the case may be) into movement. It is the equivalent of the mechanical structure in an articulated machine.

2. The *sensing operator* receives data from the mechanical operator (internal data or knowledge which will be described later) and above all data from the external world using the senses. It can, but does not necessarily, carry out localized pre-processing of data that it transmits to the brain.
3. The *decision making operator* has two inputs: one from the sensing operator and one which corresponds to the operator's own knowledge of what has to be achieved in the slave environment (software).

To sum up, data is fed in through the sensors and/or decision making operator and emerges through the mechanical (physical) operator to generate a force.

2.1.2.2 CIRCULATION OF DATA, DECISIONS AND ACTIONS [see Figure 2.1.2.(3)]

The sensing operator receives two types of data:

1. *remote data* derived from viewing the master system, the master environment, the slave system and the slave environment through the window in the shielding. (Data on overheating will be ignored here.) To this must be added sound data (transmitted by loudspeakers);
2. *kinaesthetic data* resulting from the operator's physical contact with the master arm, which keeps the operator informed on its

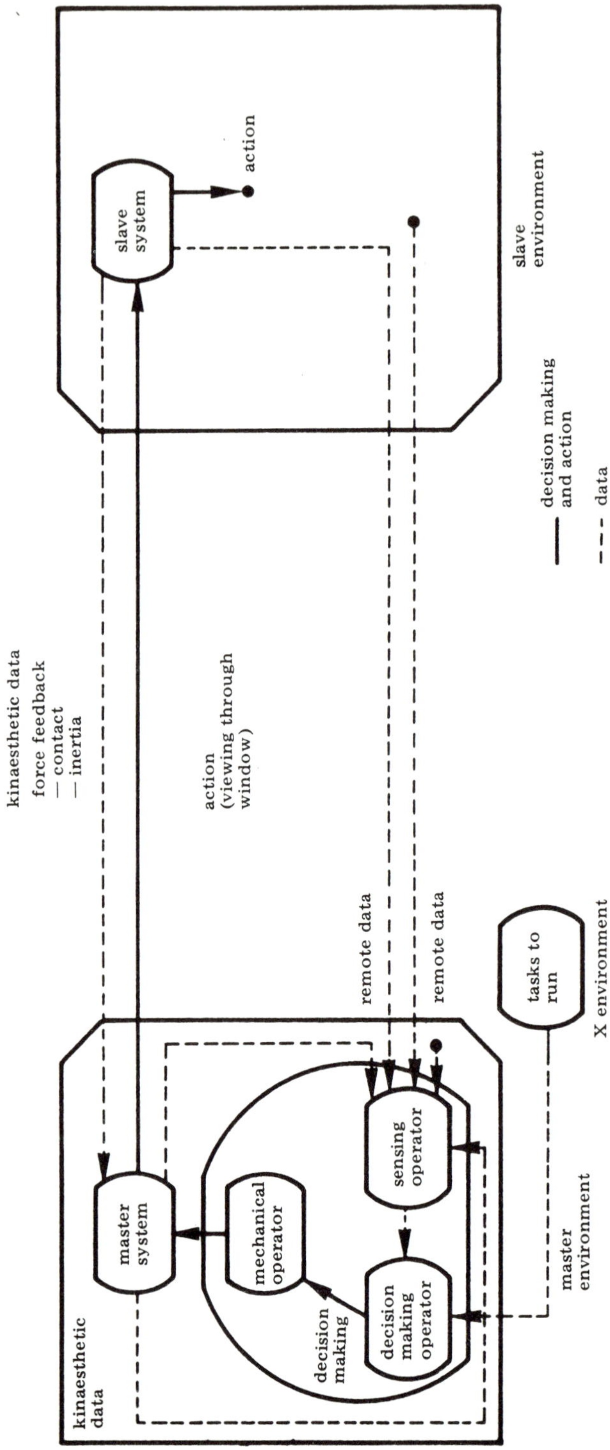

Figure 2.1.2.(3) *Mechanical remote manipulation: circulation of data, decisions and actions*

behaviour as well as on the behaviour and forces of the slave system, transmitted by the mechanical connection linking the master and slave (the pertinence of data from this last connection depends very much on its nature).

Decisions are generated by the brain on the basis of *a priori* data (a knowledge of the tasks and of the system) and sensing data. The mechanical operator transmits the action which has been decided upon to the master system, which passes it onto the slave which, in turn, executes the action.

Thus, mechanical remote manipulation is already very complicated to model, as it presupposes the knowledge of the three human functions (mechanical, sensing and decision making) which are, in fact, interdependent. In particular, performance will depend on a greater number of factors associated not only with the mechanical system used, but also with the envisaged task and the nature and quality of the data assimilated, and with the operator.

2.1.3 Teleoperation with electrical transmission (unilateral control or bilateral servocontrol)

In this type of teleoperation, the mechanical link between the master and slave systems is replaced by an electrical link using a cable or radio control (this requires the placing of actuators on the slave system) (see Chapters 1.4 and 1.5). Thus, it is possible to increase the physical space separating the master and slave environments until the point at which the data transmission delay is too great to be ignored.

Under these conditions, the operator can no longer pick up data on the slave environment immediately by direct viewing, and force feedback is not possible because of the lack of mechanical connections. Vision can be partially restored by placing television cameras in the slave environment, and bilateral servo manipulators reflect an artificial synthesis of the forces based on the measurement of the torques present in the slave arm.

The description of the system is, therefore, similar to that of the mechanical teleoperator. While having the advantage of the increased independence of the two environments (master and slave), general performance can be reduced because there is no replacement for a direct view of the slave environment; an image on a screen (even if it is stereoscopic) can never be as effective.

We are here assuming that the television circuit limits the performance of the operator-manipulator-television system. In reality, the performance of any one of these factors restricts overall performance. However, if the mechanical performance of the manipulator limits the system, a slightly inferior quality of vision will not give rise to a reduced performance. In particular, the loss of quality of vision, for want of constant data, can affect either the frequency of repetition of the images (Ranadive and Sheridan, 1981) or their steadiness, and under given conditions both must be produced to a satisfactory standard. This is particularly important when the transmission delay interferes with the overall system. This topic will be dealt with fully in Part 3.

2.1.4 Computerized teleoperation

To make up for the short-comings of electrically transmitted teleoperation, a 'second operator' can be introduced – a very limited one, but with the advantage that it can be distributed spatially. It, too, has a brain (the power to calculate, to reflect and to make decisions), artificial senses and a means of action. This is made possible by computer science. There can now be teleoperation 'with two brains' (the operator and the computer) which, in theory, should improve performance but also increases the complexity and introduces new problems, such as a possible conflict arising between the two brains.

2.1.4.1 THE ELEMENTS PRESENT [see Figure 2.1.4.(1)]

Compared with mechanical teleoperation, the master environment in this system contains, over and above the substructures already present, sensing devices connected to the computer which enable it to:

1. observe the master environment;
2. observe the master system;
3. take readings of certain measurements from the physical or mechanical operator (eg to assess the degree of physical or nervous fatigue).

To avoid overloading this description, it will be asumed that all of these sensors are passive, that is to say that they constantly monitor the specified values but that they are never adjusted. (This is not always the case but it does not affect the understanding of the phenomena.)

Similarly, within the slave environment sensors will be found which allow data readings on the slave system and environment to be taken.

But the most important modifications will be in the X environment. As before, the tasks to be accomplished will be placed there, together with the computer. Added to this could be the computer function software, comprising the operator programs and all those connected with the application. These can be included, as in the case of the operator, in the three following functional groups [see Figure 2.1.4.(2)].

1. The *physical computer* which receives its commands from two

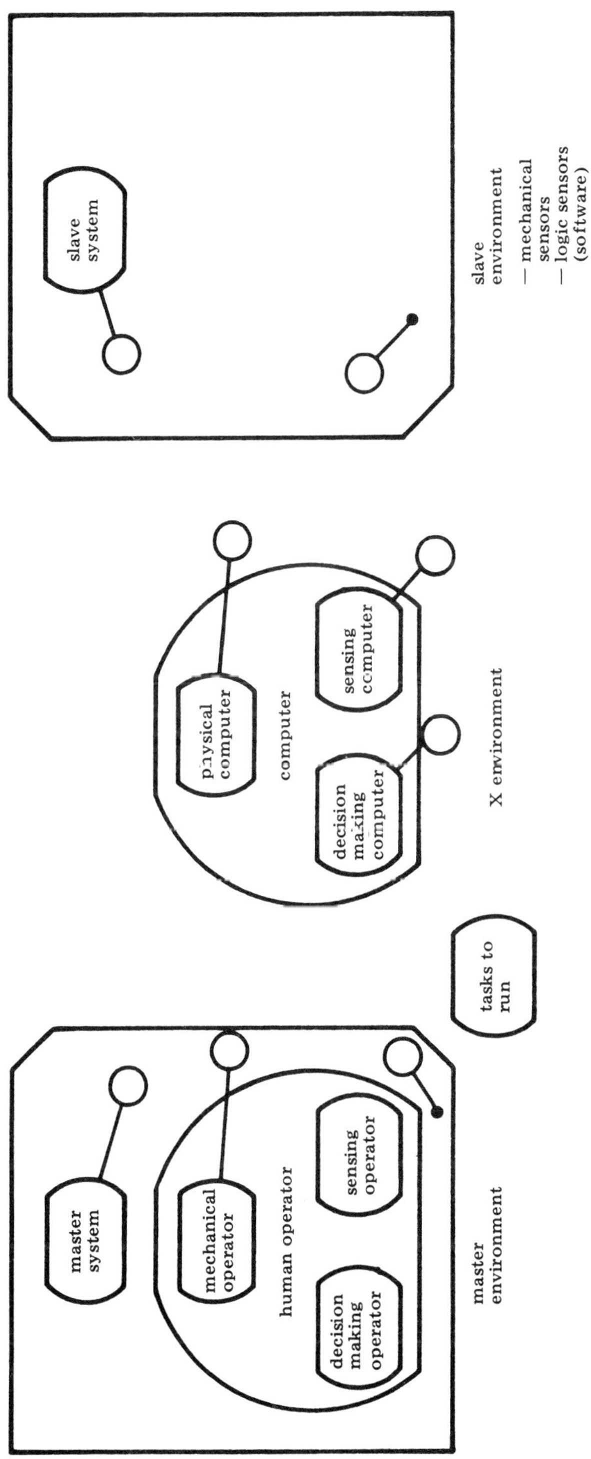

Figure 2.1.4.(1) *Computerized teleoperation*

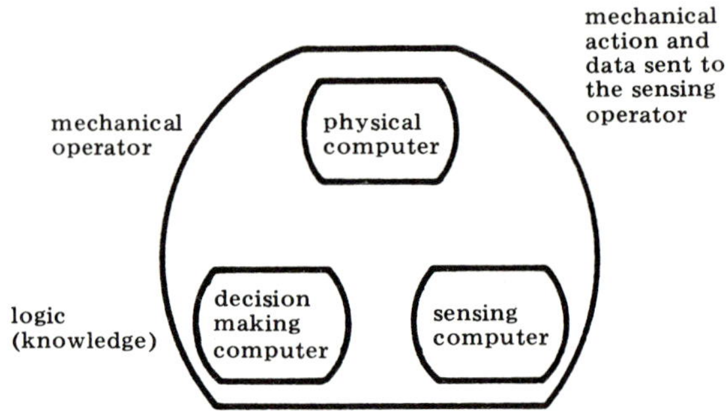

Figure 2.1.4.(2) *Three functional groups of computerized teleoperation*

sources (this is what distinguishes it, on the strategic level, from the mechanical operator in the case under consideration):
- the part of the computer in which instructions are developed (decision making computer);
- the physical operator who physically acts on a certain number of parts of the computer in response to instructions issued by the decision making operator.

The output from the physical computer corresponds to actions which are to act on the various components (master and slave system, master and slave environment – all of which in turn will exert an influence on the sensing operator). This physical computer is itself equipped with an observation system which allows it to detect any hardware failures or other relevant modifications.

2. The *sensing computer* receives data from the artificial sensors placed in the environments. It should be noted that, as was the case for the operator, this arbitrary division of the whole computer made here gives the impression that the sensing computer is nothing but software (which has the physical computer as a back-up). The sensors C′ [see Figure 2.1.4.(1)] which are joined to it are, therefore, programs (or software sensors) and make it possible for the software to be monitored.

   The output from the sensing computer is intended purely for the decision making computer.
3. The *decision making computer* acts solely on the physical computer and only receives data from the sensing computer and from software already present (eg knowledge of tasks).

2.1.4.2 DATA PATHS [see Figure 2.1.4.(3)]

Therefore, two centres of action, two centres of observation and two centres of decision making can be identified.

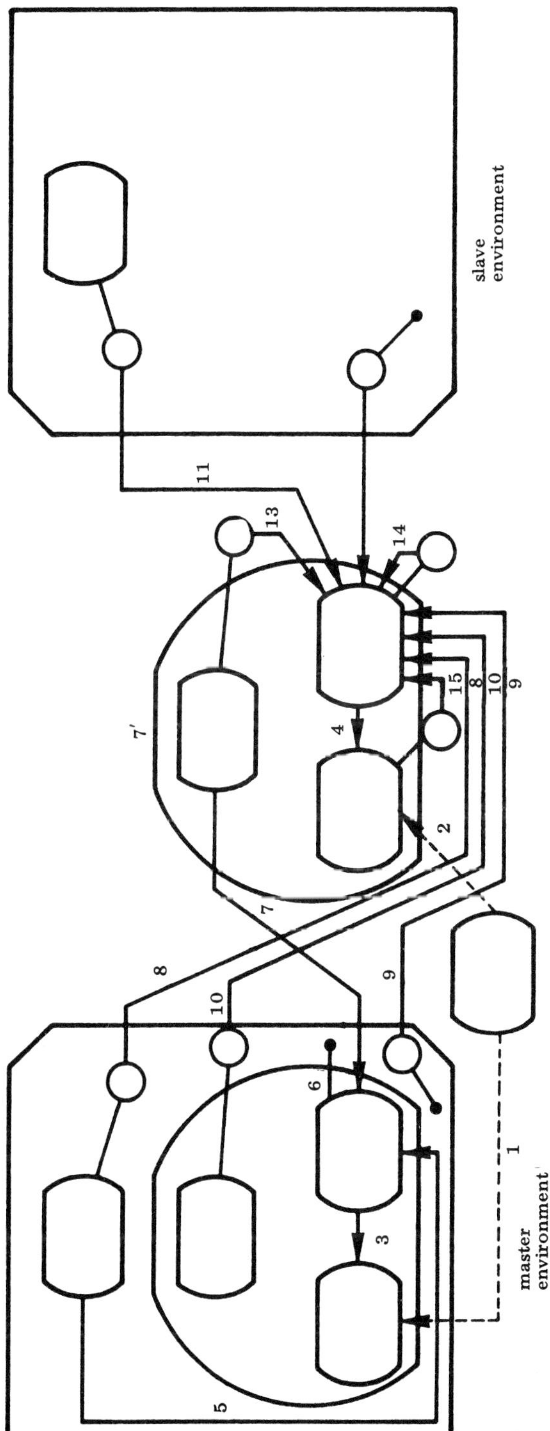

Figure 2.1.4.(3) *Data paths in computerized teleoperation*

These last two centres receive two flows of data. The first reassembles the knowledge of the tasks from the operator (arrow 1) and from the computer (arrow 2). The second flow proceeds from the sensing operator (arrow 3) and the sensing computer (arrow 4).

The sensing operator is sensitive to the master system (arrow 5) and the master environment (arrow 6) which the operator can touch and see, but equally to the physical computer which presents data to the operator via the appropriate physical output system (eg screen, tactile stimulation), which, in certain cases, the operator can touch and see (arrow 7). Vision can be by direct viewing, thus bypassing the computer (arrow 7′).

The sensing computer receives the observations of all the artificial sensors situated:

- on the master system (arrow 8);
- in the master environment (arrow 9);
- on the physical operator (arrow 10);
- on the slave system (arrow 11);
- in the slave environment (arrow 12);
- on the physical computer (arrow 13);
- on the software sensors (arrows 12 and 14).

This data flow from the multiform sensors, as yet unmentioned, suggests the complexity of computerized teleoperation.

2.1.4.3. PATHS OF DECISION MAKING AND ACTIONS [see Figure 2.1.4.(4)]

Commands are generated by the two decision making centres (decision making operator and decision making computer) based on the *a priori* data (arrows 1 and 2) and on the data supplied by the sensing stages (arrows 3 and 4). These commands are taken as signals of sufficient strength addressed exclusively and respectively to the mechanical operator and the physical computer (arrows 16 and 17).

The mechanical operator can then act on the master system (arrow 18), on the master environment (arrow 19) or on the physical computer (arrow 20). The physical computer can act on the master and slave systems (arrows 21 and 22) and on the master and slave environments (arrows 23 and 24) but also on the operator's senses (arrow 7) (informative action). The master system acts on the slave system (arrow 25) which modifies the slave environment (arrow 26). The feedback from the slave environment to the master environment can only be effected by means of the artificial sensors and the sensing computer.

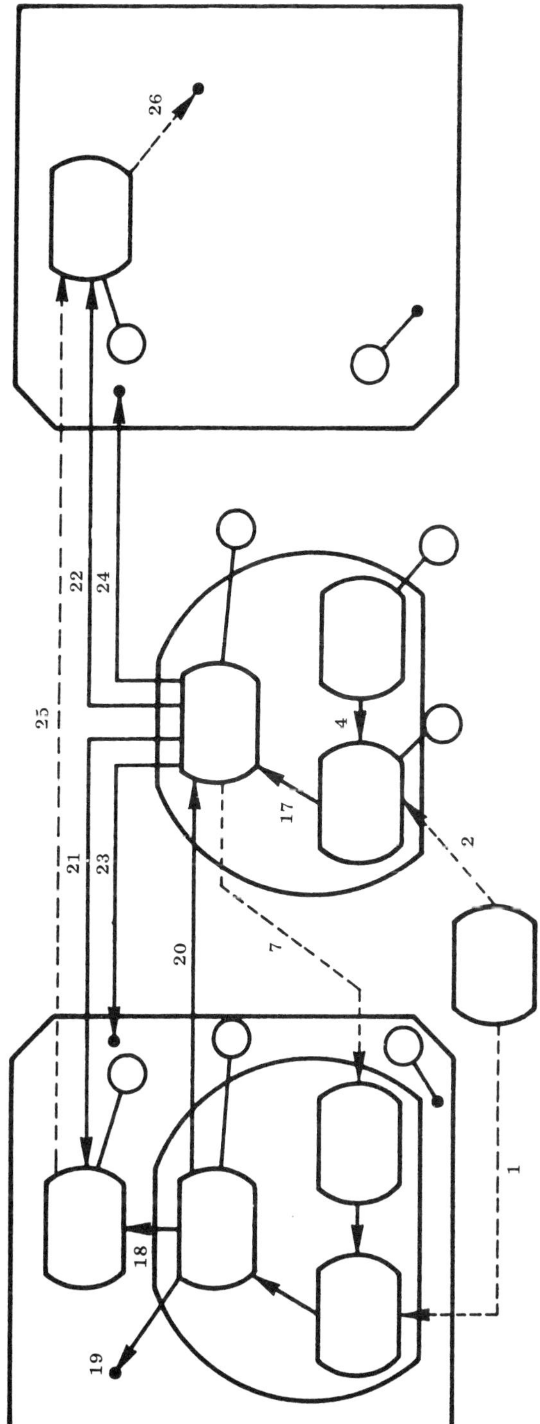

Figure 2.1.4.(4) *Decision making and action paths in computerized teleoperation*

2.1.5 Comments on this description

This conceptual description of a teleoperation system has two aims:

1. to explain as simply as possible the elements involved in a teleoperation system and their overall relationships;
2. to draw attention to the importance and complexity of these relationships, in such a way that the individual trying to model or optimize a small part of one of the elements present will bear in mind that a local improvement can diminish the overall performance of the system. In other words, in all work on teleoperation, it is important to be aware that numerous hypotheses whose validities are uncertain are being made.

Thus, the problem has been broken down into a limited number of elements. However, it is still difficult to write about these different elements in real terms, and it is more difficult still to devise a perfect system with so many parameters and unknowns. This is why, as is always the case when tackling a highly complex problem, different approaches are possible. The more detailed the study becomes, the more choices must be made, generally heuristic ones, and this involves a risk of distorting the solution to the general problem. By way of an example, a fairly detailed type of structure for an operator-robot system is shown in Figure 2.1.5.(1) (Malvache, 1981). It has the advantage that it provides, within each rectangle, either workable hardware or software for each real case, but certain structural relationships between the rectangles are chosen *a priori*.

2.1.6 The overall function of data processing in teleoperation

In standard (ie mechanical) teleoperation, it can be said that 'the operator does everything'. The operator is familiar with the appointed tasks, has a certain knowledge of how far advanced the task is and is constantly developing a plan of action on the master components to make the task in view progress properly.

In addition, the operator provides all the mechanical power necessary for the completion of the task, including that lost through faults in the telemanipulator. In the case of teleoperation using a bilateral servo manipulator, the force feedback ratio, which can be less than one, divides the force applied by the operator and thus his fatigue by this amount. This is similar to the situation of a driver and car or a pilot and aeroplane, where the amount of power needed to steer it becomes smaller and smaller in relation to that provided by the machine.

To achieve this the operator uses his knowledge of the machine, his experience and his initiative.

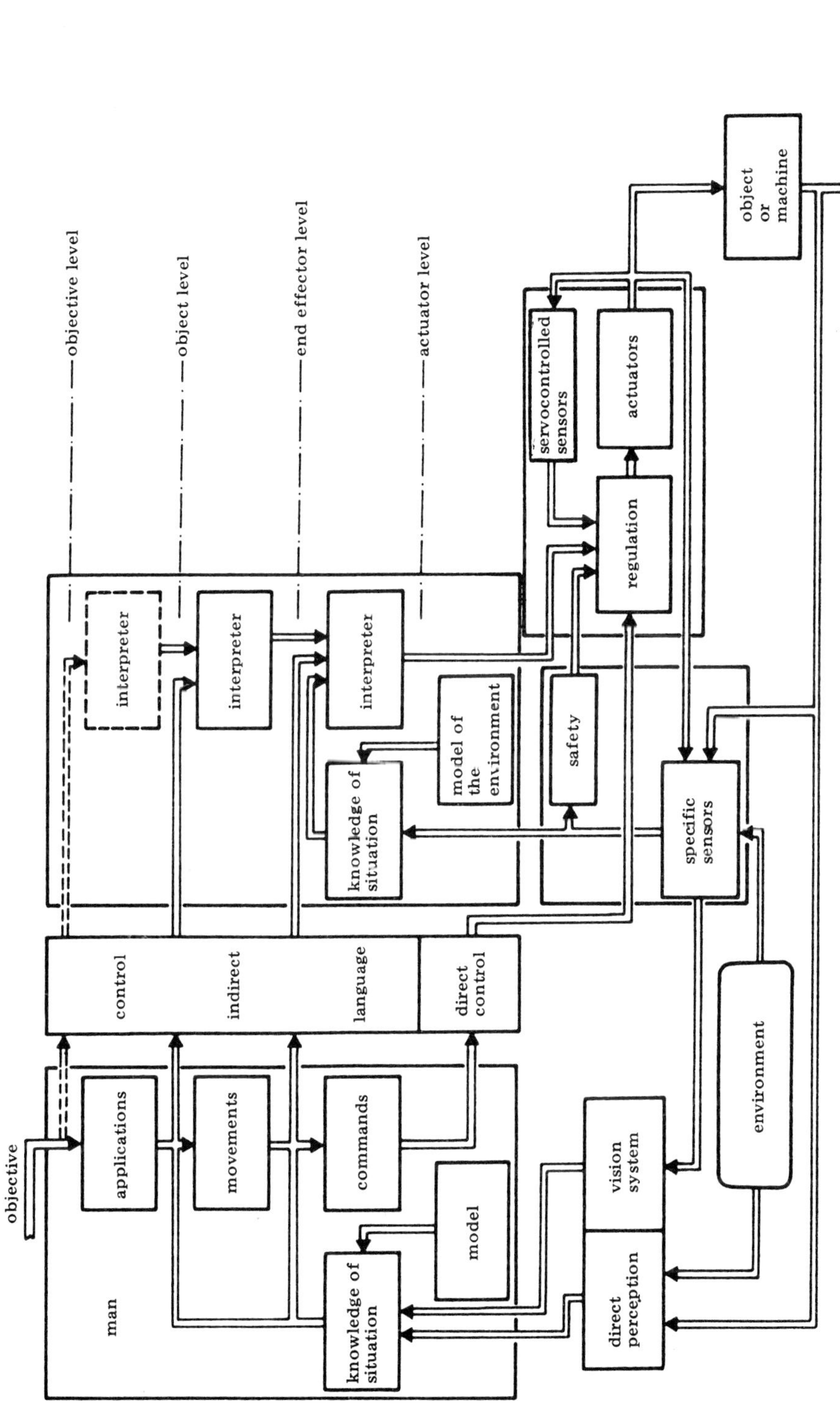

Figure 2.1.5.(1) *Detailed structure of the operator-robot system (from Malvache, 1981)*

In an ideal situation, if the machine were perfect, remote manipulation, after a short period of training, should be as if the operator were performing the task bare handed and not manipulating distant gripping tools. The operator would then be in a state of *telesymbiosis* with the hostile environment where the slave system is located (Charles and Vertut, 1974).

It is doubtful whether this can ever be achieved, but data processing makes it possible to approach this state of ideal 'transparency' by considerably relieving the physical and psychological strain of work on the operator.

Of course, it can also be argued that the addition of the computer should represent such a great advance that it is no longer necessary to aim at removing obstacles between the operator and the task, but rather simply to substitute the computer for the human brain. The master component and the human operator would thus no longer need to exist. Instead the mechanism would constitute a real autonomous robot comprising the slave system and the computer. In reality, however, teleoperation tasks are so complex and so varied that for the time being this hypothesis cannot be considered.

The computer is used to improve the teleoperation system and is seen as an aid to the entities in the system, in particular to the human operator.

This help (the robotic aspect of teleoperation) can manifest itself in three ways, which are complementary and sometimes overlapping.

2.1.6.1 THE USE OF DATA

The task is to help the operator understand more clearly what is happening (ie to perceive more clearly the state of the system and of the environment). The computer can provide a considerable amount of useful data in a form accessible to the operator's senses:

- reconstitution of touch or the sense of force;
- generation of sounds or words to indicate a certain danger or a certain significant phenomenon;
- reconstitution on the screen of the slave environment;
- data in the form of drawings, graphics, text etc.

The choice of presentation of parameters to be communicated to the operator is very varied and depends greatly on the objective in view. The aim is to help the operator to develop a plan of action by supplying data which would be inaccessible without the sensors and the computer, either because they are physically unattainable (eg seeing the slave universe) or because of the interference present (eg force feedback).

2.1.6.2 TYPES OF OPERATOR SUBSTITUTION

First, imagine that in the course of a task to be carried out by teleoperation, a vessel containing a fluid must be transported. It would be desirable to maintain the gripper in a horizontal position throughout the transportation. This could pose problems given a master station with an articulated system of six DOF because of the potential mechanical linkage of the DOF on the slave arm and also because it is difficult to ensure a change of position of the gripper without introducing some rotation.

Thus, the problem of coordinating the articulation so as to accomplish the objective arises. The computer can ensure this coordination and maintain the gripper in the horizontal position automatically, whatever actions the operator may take on the DOF. The computer has replaced the operator by controlling orientation in parallel.

Other examples where the responsibility for the DOF has been taken over have also been explored at Saclay (eg holding a transformer disc tool in its plane and a drill on its axis).

Another example involves the elimination of weight on the object held by the remote manipulator. The computer in this case simulates the effects of gravity on the object and subtracts this from the kinaesthetic feedback to the operator. In this case, in a general way, the computer analyses data transmitted by the slave assembly and supplies it in an improved form to the master system and then to the operator.

Substitution can also be considered, this time a partial substitution in the time involved and not in the number of functions performed. It can involve, as in the case described by Sheridan, a largely repetitive programmed activity which the operator can introduce into the execution of the overall task, and during the execution of the task the operator intervenes only to supervize. This task can be more or less self-adaptive depending on the degree of automation of the system in question.

This is seen when an object is grasped automatically once the operator has guided the sensitive gripper to the object. The operator then activates the 'automatic grasp' function and, with the help of ad hoc sensors, the computer takes over from the operator in the final phase of the grasping operation. For this type of performance of subtasks, then, the system works like a robot [see Figure 2.2.2.(2)].

In these different examples, three types of substitution can be seen: one which introduces into the system the equivalent of the operator's reflexes, thus eliminating excessive tiredness or risk of operator error, another which involves a filtering of data and the improvement of its presentation, and one which introduces total automation, though of limited duration, into the execution of the task.

2.1.6.3 THE PREVENTION AND SAFETY FUNCTION

The task of the computer is to prevent the operator from performing actions whose results could impair the life expectancy or durability of the hardware. This preventive function can take the form of passing on data: the computer warns the operator that the manipulator is approaching its limits, or is about to collide with an obstacle, or that an engine component is rubbing dangerously etc [see trajectory 7 in Figure 2.1.6.(1)]. It can take the form of active intervention: the computer quite simply prevents certain manoeuvres by blocking the relevant instruction [see trajectory 22 in Figure 2.1.6.(1)]. In other words, it takes on the function of automatic surveillance over the totality of entities present in the slave and master environments, and a decision making power.

2.1.6.4 RELATIONSHIP BETWEEN THE OPERATOR AND THE COMPUTER

In computerized teleoperation, the presence of two decision making centres (the operator and the computer) can lead to conflict, despite the ultimate aim of total collaboration.

To avoid such conflicts, there are two possible approaches based on two diametrically opposed hypotheses. The first proceeds from the supposition that the operator is so limited, clumsy and absent minded that whilst the operator must accept a certain weight of work in the operation of a remote manipulation system, the highest degree of safety and efficiency can be achieved by entrusting the top level of supervision to the computer, which is not susceptible to fatigue or distraction and is thought of as being faultless.

The second hypothesis, on the other hand, proceeds from the supposition that the operator, who is eminently well adapted, only requires that the computer should act in very particular cases and more to provide data rather than to perform actions.

It can be said that the respective responsibilities of the operator and the computer are best defined by a compromise between these two attitudes, a compromise dictated by the potential consequences of an action by the operator or the computer. For example, the computer can force the operator to avoid an obstacle, but the operator must be able to override this function if, in the course of a manoeuvre, he wants to get closer to the obstacle (which has become a point of interest) than the computer will allow. It is clear that in the last resort the operator must be able to overrule the computer.

In practice, this dual control of the system by operator and computer does not yet seem to have been the subject of far-reaching research (Vertut *et al.*, 1984I, II, III; Roger and Millot, 1980) and, in reality, data processing only has a role in the performance of highly specific commands (Quetin, 1978).

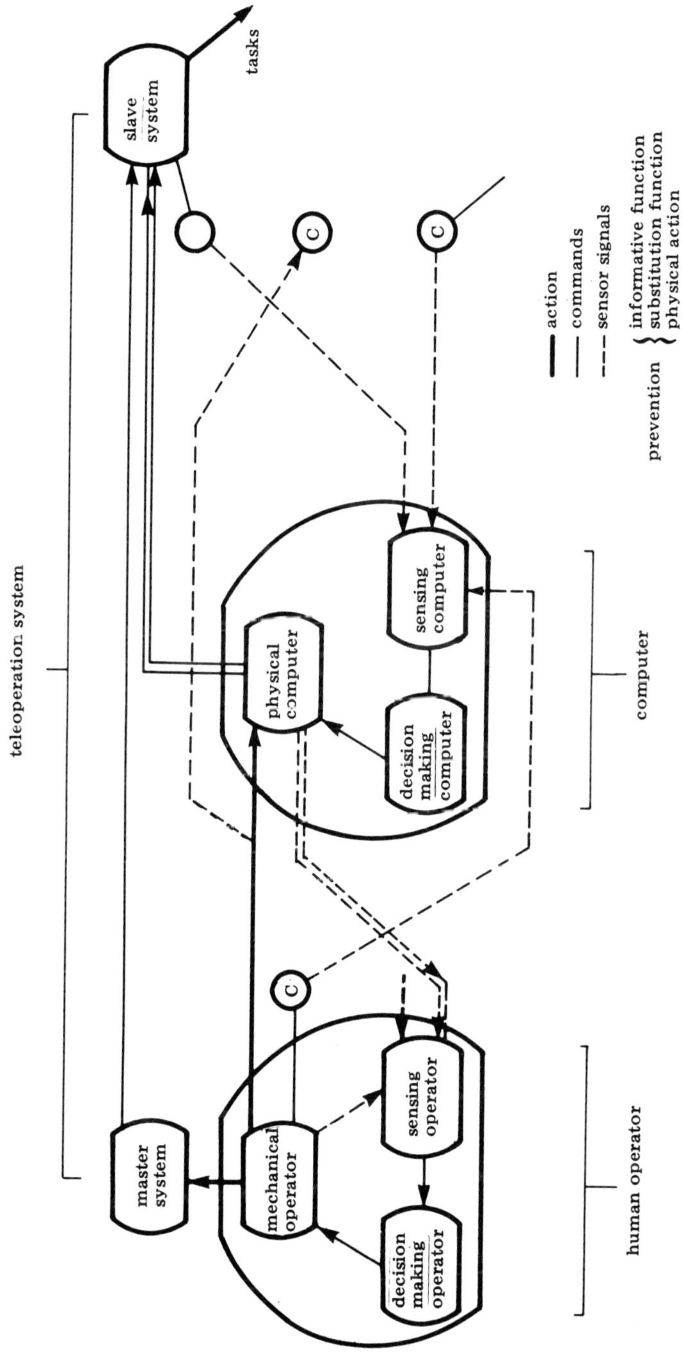

Figure 2.1.6.(1) *The functions of the computer in the teleoperation system*

It should be noted, in passing, that the task of surveillance which devolves onto the computer can be applied, as has already been indicated, not only to the teleoperation system and its environment, but also to the operator and the man-machine interface (eg the detection of fatigue and the essential stopping of the system). Figure 2.1.6.(1) shows the functions of the computer. These relative roles can vary greatly according to whether the application involves standard remote manipulation or the movement of a large transporter or of a vehicle.

Conclusions

This chapter has tried to show the complexity of teleoperation systems, which results in low levels of efficiency, especially in systems controlled by computer. The approach adopted throughout this text is to describe the problems that arise at the operational level. Artificial intelligence is not dealt with in detail (Ferrell and Sheridan, 1967) (see Volume 5 of this series entitled *Logic and Programming*) but it is considered merely as an arithmetical technique to help solve localized problems which require analysis (eg helping the human operator to make decisions or to recognize shapes).

In this volume, then, some practical implications of the two main functions of the computer in teleoperation will be examined. These functions are keeping the human operator informed and replacing the human operator in some tasks.

Chapter 2.2

The operator substitution function by computer

2.2.1 Introduction

This role fulfilled by the computer, mentioned in Chapter 2.1, entails the computer momentarily replacing the operator in the control of entities present in the master and slave environments, with the exception of the operator himself. The implementation of this substitution function implies, first of all, a strategic aspect to determine in what conditions and in what manner control will be handed over to the computer. If the preventive aspect of the function in question is excluded from consideration, there is a general rule that the operator instructs the computer to take over from him in certain tasks or subtasks. The commands are transmitted through the normal input channels of the computer (eg switches, teletype units, vocoders).

Research into the automatic detection of conditions in which the computer takes over from the human operator, and vice versa, has not yet reached an advanced level (Jolly, 1982; Freedy and Weltam, 1972; Freedy *et al.*, 1971; Corliss and Johnson, 1968; Sheridan and Ferrell, 1974). By simplifying the issue somewhat, the computer can be said to intervene in three main ways, when the substitution function is activated:

1. In the first mode, the human operator is eliminated. The slave system acts with the aid of the computer like a real robot (see Volume 2 of this series entitled *Interaction with the Environment: Robot Sensors and Sensing*).
2. In the second mode, the operator and the computer work in parallel, acting on different systems which, in principle, are not mechanically coupled to each other. Therefore, when the operator manoeuvres the master system, and consequently the slave system, the computer will supervize the manoeuvres and modify the elements in the master and/or slave systems which are not part of the master-slave system itself (eg automatic tracking of cameras).
3. In the third mode, the computer modifies the normal transmission of the movements and forces between the master and slave parts.

This chapter gives a number of examples of these modes of computer intervention.

2.2.2 Elimination of the human operator upon request

For the computer-slave system unit, this consists of carrying out a certain number of tasks in automatic mode. This unit effectively constitutes a robot, and it should be noted that it must belong to the second generation at least,* that is, it can adapt to its environment by itself. At this point it should be noted that a characteristic which is almost always present in teleoperation is the geometrical non-repetitiveness of the operations to be carried out, and their relative unpredictability. The consequences of this are that:

1. only the subtasks within a macrotask can be automated [these subtasks are encountered in many different macrotasks (see Volume 2 of this series entitled *Interaction with the Environment: Robot Sensors and Sensing*)];
2. actions can be carried out only in adaptive mode with respect to the environment;
3. the process of automatic subtask execution must be initiated by the human operator.

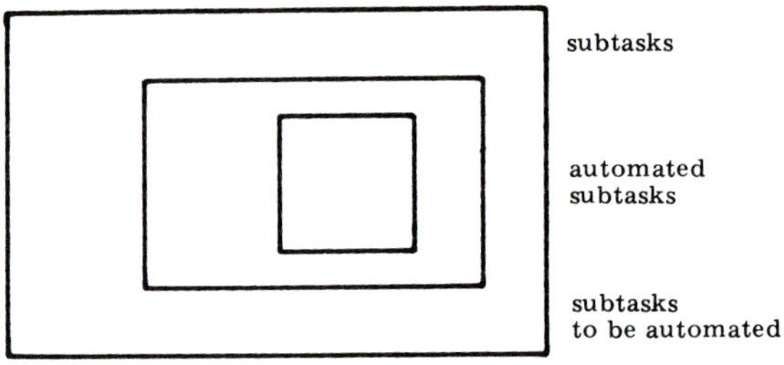

Figure 2.2.2.(1) *Automation of subtasks to improve performance*

Clearly, the most useful subtasks to automate are those that improve the performance of the teleoperation system to a considerable degree, because of the difficulty in their direct execution by the operator [see Figure 2.2.2.(1)]. In remote manipulation, it is generally difficult to obtain a satisfactory level of positional and orientational precision for the terminal device with respect to the environment. This precision is often necessary to allow grasping of certain objects, or precise seam or

*Since the appearance of the first volume in this series, a new classification for robot generations has been developed. First generation: robots functioning in an open loop with respect to their environment; second generation: robots able to adapt themselves to certain variations in the environment with the use of external sensors; third generation: robots making extensive use of artificial intelligence techniques in order to comprehend and decide on courses of action.

line tracking (eg for sanding, trimming, sawing, drilling) or assembling parts (eg inserting and tightening a screw).

Execution strategy of a macrotask can be described as: the operator moves the slave arm by moving the master arm, which can involve different movements of great amplitude and little precision (eg to guide the gripper to the object of interest). It falls on the computer to run the more precise subtasks (eg to grip the object correctly) and allows the operator again to control the slave arm.

When automating subtasks, let us suppose, as already shown, that the slave system is equipped with external sensors and that the computer is aware of the subtask to be run and the strategy of use of the sensors. This general problem is described in Volume 2 of this series entitled *Interaction with the Environment: Robot Sensors and Sensing.*

Figure 2.2.2.(2) illustrates the types and uses of the sensors placed on the gripper of the teleoperator (Bejczy, 1979; André, 1984).

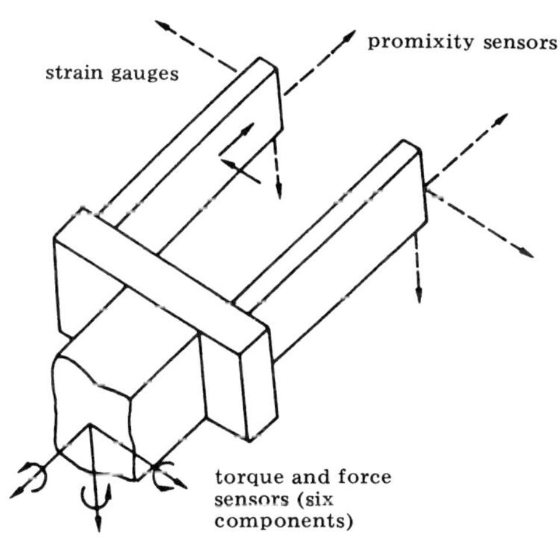

sensors	data generated	use
proximity sensors	presence data for the space between the gripper and the object	to locate object(s) by adjusting the alignment of the gripper and to avoid obstacles
strain gauges	distribution and strength of pressure between the gripper and the object	to effect contact, to push, to grip and to adjust the gripping force
torque or force sensors	strength of forces and torques exerted by the gripper with respect to the set of coordinate axes	to move the objects steadily without jerking

Figure 2.2.2.(2) *The types and uses of the sensors fitted to the teleoperator gripper: IRISA advanced gripper equipped with proximity sensors (ARA project-La Calhène)*

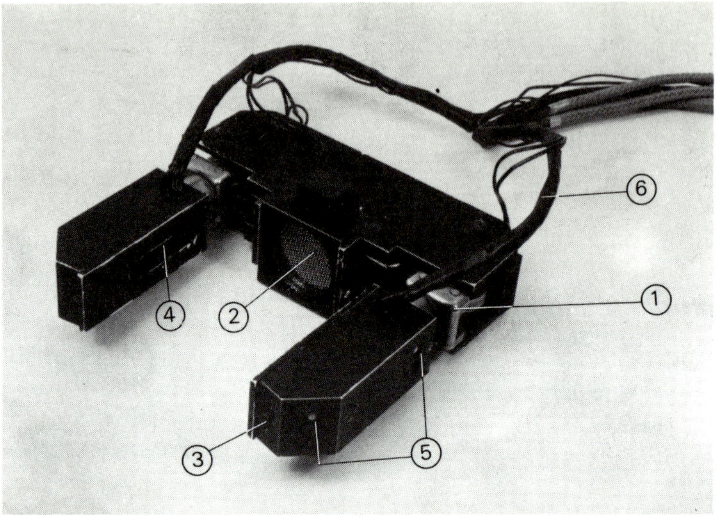

Figure 2.2.2.(3) *Gripper equipped with proximity sensors (IRISA-CEA-La Calhène, 1982);* 1, *body and fingers of the CRL gripper, which has a gripping force of 50 daN;* 2, *ultrasonic sensors which supply the correct distance between the object to be gripped and the gripper;* 3, *infrared proximity sensors for centring the object;* 4, *infrared range for the end of gripping;* 5, *infrared proximity sensors to avoid obstacles (bench and shielding);* 6, *relationship between electronics and pre-processing [see Figure 1.7.10.(1)]. Two sensors are required for centring on the object. Three sensors are required for tracking a parallel surface. A fourth sensor is required for orientation. Photo courtesy of IRISA*

The other signals generated by the sensors are sent to the two decision making centres (the operator and the computer) but they are used in different ways: the computer's strategy is preset; the operator's strategy is worked out by him in his head.

2.2.3 Parallel action of the operator and the computer on different systems to ensure their coupling

Another possible type of computer action consists of overseeing the human operator and/or the master and slave systems so as to allow control of the other objects in the environment, and on-line adaptation of them to the needs of the system at each instant.

It is, therefore, possible to imagine the automatic modification of the position of certain objects in the master environment, as a function of the work being carried out (eg adjustment of the operator's seat, or mobility of the display screens and other modules involved in control). In the slave environment, an adaptation in the type of support of the objects of interest could be envisaged, as a function of the configuration of the slave arm.

This corresponds, in particular, to the adaptation of the large transporter of the slave arm, allowing the arm to be placed in the optimal configuration, automatically. Another example is based on the use of a conveyor belt to move objects into the optimum zone from the point of view of their easiness to be grasped.

An interesting property that is fairly independent of the nature of the tasks is the mobility of some of the sensors in the slave environment, particularly the remote data sensors and specifically the cameras. A type of application involving these techniques, already mentioned in Sections 1.1.4 and 1.7.3, is described in Section 2.2.3.1.

2.2.3.1 SERVOCONTROL OF A MOBILE CAMERA IN THE WORK ZONE OF A MANIPULATOR

One of the difficulties of teleoperation lies in the use of vision for the operator in monitoring what is happening in the slave environment. Despite the presence of mobile television cameras in this environment, it is not unusual for the zone of interest (usually around the gripper, the tool held in the gripper, and the object on which the system is acting) to lie outside the field of the camera(s), and for the operator to have to spend considerable time maintaining visual contact. This is the origin of the idea of coupling the camera movements with the displacements of the slave arm, so that the camera can be servocontrolled at the gripper or the tip of the tool [see Figure 2.2.3.(1)]. A system of this type may

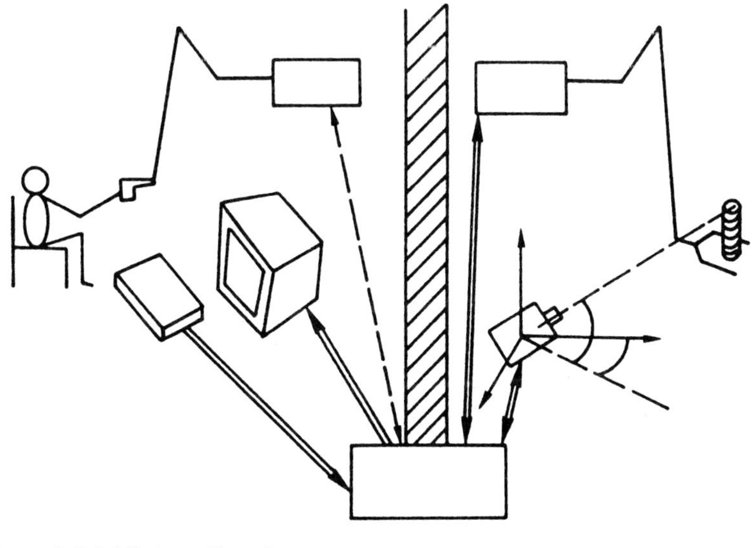

Figure 2.2.3.(1) *Coupling the camera movements with the displacements of the slave arm: C, mobile pan and tilt camera; H lies on the slave arm and is being tracked;* **1**, *computer;* **2**, *television monitor;* **3**, *teaching handle*

appear simple at first sight, but, in reality, in order for it to be effective, it must have a certain number of properties that fall outside the boundaries of standard servo-systems.

1. Strict servocontrol of the direction of camera C on point H must not be permanent. In this situation, when the gripper is constantly moving during execution of a task, the image on the screen will also move constantly, and this could inconvenience the operator (causing a sensation of seasickness).
2. If for any reason (eg during installation of a system), the image of the terminal device runs off the screen, it must be possible to relocate it rapidly and automatically.
3. Depending on the distance between the camera and the object being viewed, the image of the part of interest to the operator will appear larger or smaller. If the dimensions of the image are to be kept constant, the zoom must operate automatically.
4. In the same way, the variations in the camera-object distance will require on-line corrections to the adjustment.
5. In situations where the level of lighting varies, the aperture must be controlled automatically.

The way in which these points have been resolved in the systems described in Kuspriyanto (1981) and Vertut *et al.* (1983II) and applied successfully to various remote manipulation systems is summarized below. The cameras used are either pointed directly at the object, or indirectly via a mobile prism (since this has low inertia, its rate of response and velocity of displacement will be far more rapid than that of a much heavier video camera, moving as a block) [see Figure 2.2.3.(2)].

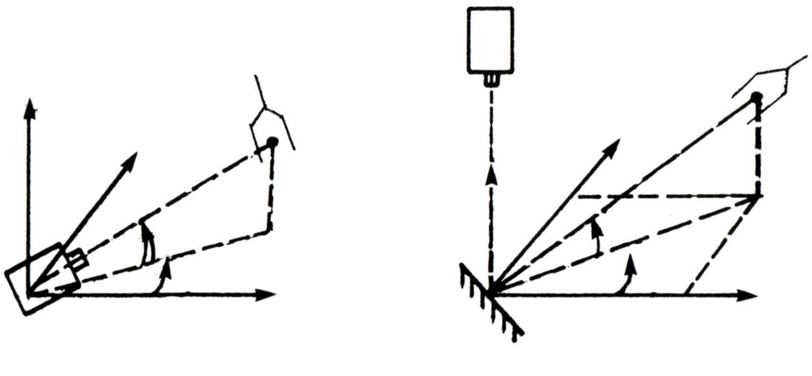

Figure 2.2.3.(2) *Position of cameras and object (H)*
(a) *Direct design: C is a mobile camera; H is the point being tracked*
(b) *Indirect design: fixed video camera; P is a mirror or mobile prism.* 1, *panning angle;* 2, *tilting angle [see Figure 1.7.4.(2)]*

Calculation of the site and azimuth (pan and tilt) of the camera and the camera-object distance

For this type of system, the first value to be established on-line concerns the direction of the camera (or the desired orientation of the prism). Consider the example of an MA23M-type (CEA-La Calhène)

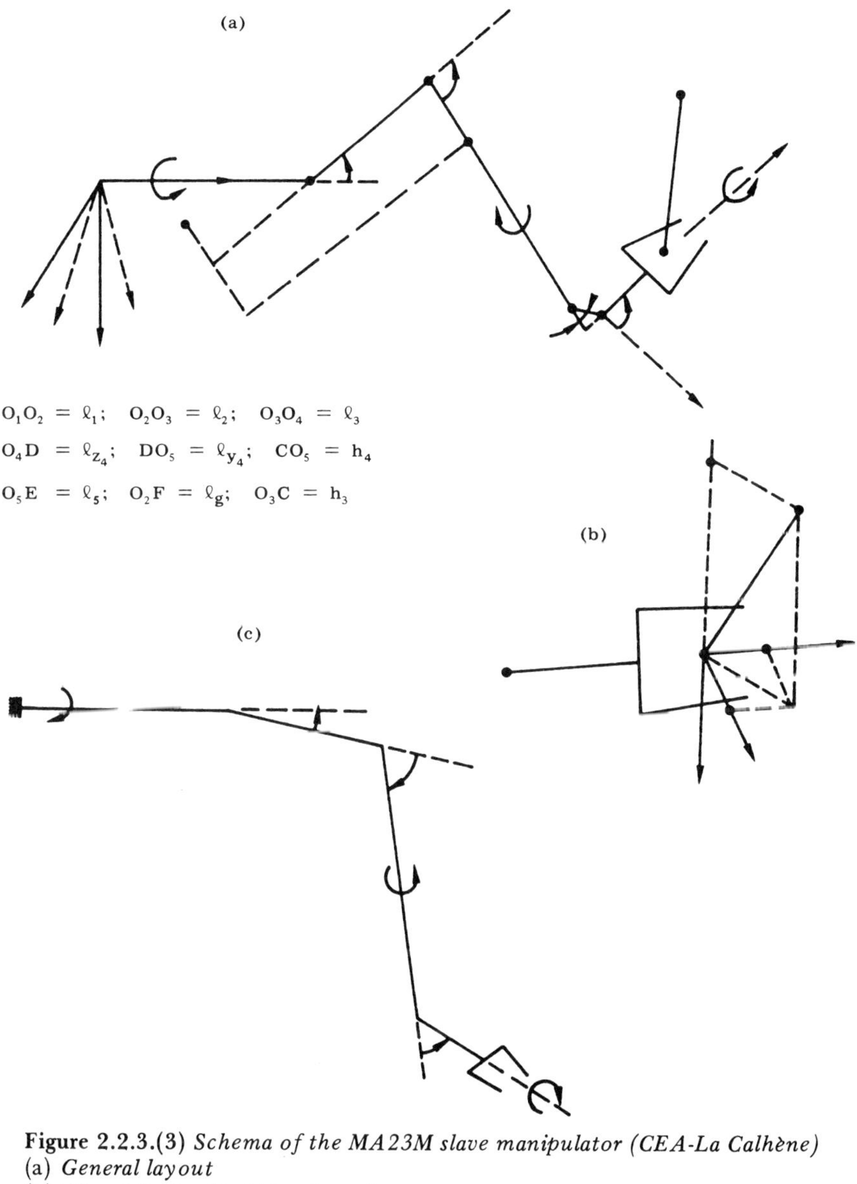

Figure 2.2.3.(3) *Schema of the MA23M slave manipulator (CEA-La Calhène)*
(a) *General layout*
(b) *Tip of the tool held by the gripper*
(c) *Configuration at the origin*

slave manipulator, the structure of which was considered in Volume 1 of this series entitled *Modelling and Control*, and the schema of which is shown in Figure 2.2.3.(3).

1. Its configuration is known (the value of the articular angles θ_i) through the readings on the potentiometers placed on the motor axes (ψ_i angles). The following equation can be established:

$$\underline{\Theta} = P_1 \underline{\Psi} + \underline{\Theta}^{\circ} \tag{2.2-1}$$

In equation (2.2-1):

$\underline{\Theta} = (\theta_1, \theta_2, \ldots \theta_6)$ is the vector of the generalized variables;
$\underline{\Theta}^{\circ} = (\theta_1^{\circ}, \theta_2^{\circ}, \ldots \theta_6^{\circ})$ is the original configuration (for which the potentiometer tensions are zero);
$\underline{\Psi} = (\psi_1, \psi_2, \ldots \psi_6)$ is the vector of the angular range of the motors;
(P_1) is the constant term matrix with dimensions 6×6, characterizing the transmissions between the motors and the articulation (assumed to be rigid).

2. The coordinates of the point of interest H (which will be in the centre of the screen) must be calculated with respect to the fixed set of coordinate axes $OX_0Y_0Z_0$. For this, let d_n, d_a and d_s be the three components of EH (E is the 'centre' of the gripper) of the set of coordinate axes associated with the gripper.

The methods of calculation given in Volume 1 of this series entitled *Modelling and Control* for the change of sets can be used here. R_0 is the fixed set of coordinate axes. The other sets are obtained as shown in Table 2.2.1.

name of set of coordinate axes	obtained by rotation of	about	of an angle	transformation matrix
R_0				fixed set of coordinate axes
R_1	R_0	Z_0	θ_1	M_0^1
R_2	R_1	X_1	θ_2	M_1^2
R_3	R_2	X_2	θ_3	M_2^3
R_4	R_3	Z_3	θ_4	M_3^4
R_5	R_4	X_4	θ_5	M_4^5
R_6	R_5	Z_5	θ_6	M_5^5

Table 2.2.1. *Sets of coordinate axes described*

For a rotation about the Z axis, the following applies:

$$M^{i}_{i-1} = \begin{pmatrix} Ci & -Si & 0 \\ Si & Ci & 0 \\ 0 & 0 & 1 \end{pmatrix} \qquad (2.2\text{-}2)$$

and for a rotation about the X axis:

$$M^{i}_{i-1} = \begin{pmatrix} 1 & 0 & 0 \\ 0 & Ci & -Si \\ 0 & Si & Ci \end{pmatrix} \qquad (2.2\text{-}3)$$

where $Ci = \cos\theta_i$ and $Si = \sin\theta_i$. Therefore, a vector $\underline{V_i}$ expressed in the set of coordinate axes R_i is expressed in the set of coordinate axes R_{i-1} by:

$$\underline{V_{i-1}} = M^{i}_{i-1}\underline{V_i} \qquad (2.2\text{-}4)$$

The coordinates of point H in R_0 will thus be given by:

$$(O_1H)_0 = (O_1E)_0 + M^6_0\,[d_s d_n d_a]^T \qquad (2.2\text{-}5)$$

$$= (O_1O_5)_0 + M^5_0\,[0 0 \ell_5]^T + M^5_0 M^6_5\,[d_s d_n d_a]^T \qquad (2.2\text{-}6)$$

$(O_1O_5)_0$ and M^5_0 can be easily obtained:

$$M^5_0 = M^1_0 M^2_1 M^3_2 M^4_3 M^5_4 \qquad (2.2\text{-}7)$$

and the coordinates of O_5 are:

$$\begin{pmatrix} X_{O_5} \\ Y_{O_5} \\ Z_{O_5} \end{pmatrix} = \begin{pmatrix} S1S2\ell_2 + S1S(2+3)k_2 + C1S4\ell_{y_4} + S1C(2+3)C4\ell_{y_4} \\ -C1S2\ell_2 - C1S(2+3)k_2 + S1S4\ell_{y_4} - C1C(2+3)C4\ell_{y_4} \\ \ell_1 + C2\ell_2 + C(2+3)k_2 \rightarrow (2+3)C4\ell_{y_4} \end{pmatrix}$$

(2.2-8)

The position of H in R_0 can, therefore, be established.

3. The camera-object distance O′H [see Figure 2.2.3.(4)] is also required. The distance O_1O' is assumed to be known and fixed, and therefore:

$$D = |\overrightarrow{O'H}| = |\overrightarrow{O'Q_1} + \overrightarrow{O_1H}| \qquad (2.2\text{-}9)$$

4. In order to find the site and azimuth (pan and tilt) of the camera (or prism) corresponding to its being pointed at H, it is assumed that the camera is fixed, and has a rotating prism placed as shown in Figure 2.2.3.(5), where $\underline{O'H_{x'}}$, $\underline{O'H_{y'}}$ and $\underline{O'H_{z'}}$ are the components of $\underline{O'H}$ in the set of coordinate axes O′X′Y′Z′, giving:

$$\text{site} = \tfrac{1}{2}\arc\sin \frac{\overline{O'H_{y'}}}{D} - \frac{\pi}{4} \tag{2.2-10}$$

$$\text{azimuth} = \arccos \frac{\overline{O'H_{z'}}}{D\cos\alpha_1} \quad \text{Si}\,\underline{O'H_{x'}} \geqslant 0 \tag{2.2-11}$$

$$= 2\pi - \arccos \frac{\overline{O'H_{z'}}}{D\cos\alpha_1} \quad \text{Si}\,\underline{O'H_{x'}} < 0 \tag{2.2-12}$$

$$\alpha_1 = \arcsin\left(\overline{O'H_{y'}}/D\right) \tag{2.2-13}$$

As a result of these calculations, we now have a series of values which allows the following to be established at any time:

1. the camera (prism)-object distance (which may be used for adjusting and enlarging);

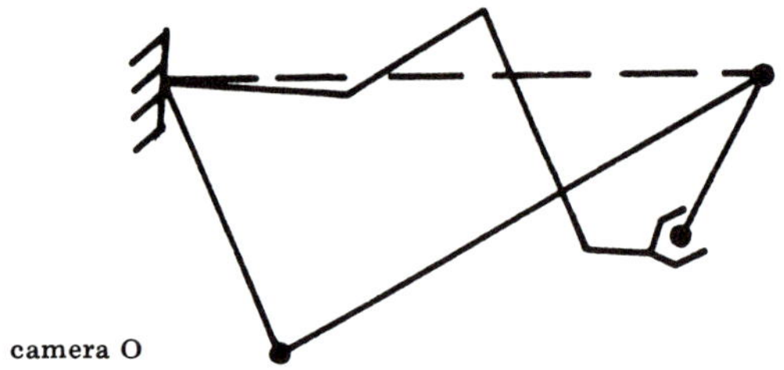

Figure 2.2.3.(4) *Camera-object distance O'H*

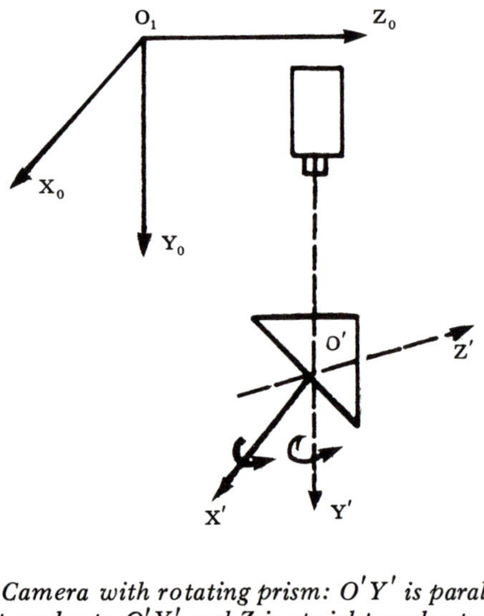

Figure 2.2.3.(5) *Camera with rotating prism: $O'Y'$ is parallel to O_1Y_0; $O'X'$ is at right angles to $O'Y'$ and Z is at right angles to the prism*

2. the angles of site and azimuth to apply to the motorized prism in order to direct it at the object H [the coordinates of which are assumed to be known in the set of coordinate axes associated with the gripper, known as shift coordinates (of H with respect to the centre of the gripper), and can be displayed at the control station. In practice, shifting is regulated by training, or programmed, depending on the tool].

Strategy for tracking an object using a camera

It has already been stated that permanent tracking of H by varying the pan and tilt can be achieved if requested by the operator. Tracking can be defined as the automatic recentring of the image of H on the screen, only when this image falls outside a certain circle (displayable radius) [see Figure 2.2.3.(6)].

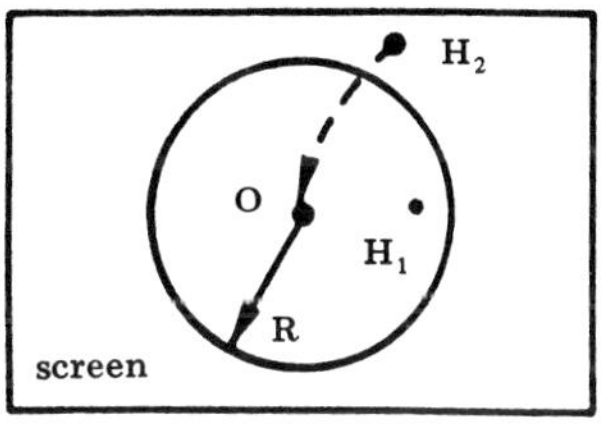

Figure 2.2.3.(6) *Tracking an object using a camera. If H is positioned at H_1, for example, inside the circle, the camera cannot move. If H lies outside the circle at H_2, for example, the camera can film the object*

To define the conditions for realignment, it is assumed that the camera's field of vision is conical [see Figure 2.2.3.(7)]. After a simple displacement of the gripper, H is located at H′. Note that:

ΔSi = 2[site (H′) − site (H)] (if a mirror is used), and
ΔAz = azimuth (H′) − azimuth (H).

A circle of radius r, in the plane of the camera image appears as a circle with radius r′ on the VDU screen. The factor of camera-screen amplification is G = r′/r. The circle of radius R situated in the field of vision appears on the screen with surface $S = \pi r^2 G_c^2$. On the VDU screen, the circle corresponding to that with radius H′K in the field of vision has, as its surface:

$$S' = \pi D^2(\Delta Si^2 + \Delta Az^2)G_c^2 \frac{f^2}{(D-f)^2} \qquad (2.2\text{-}14)$$

If the ratio S′/kS = N is considered, then:

$$N = f^2(\Delta Si^2 + \Delta Az^2)kr^2 \qquad (2.2\text{-}15)$$

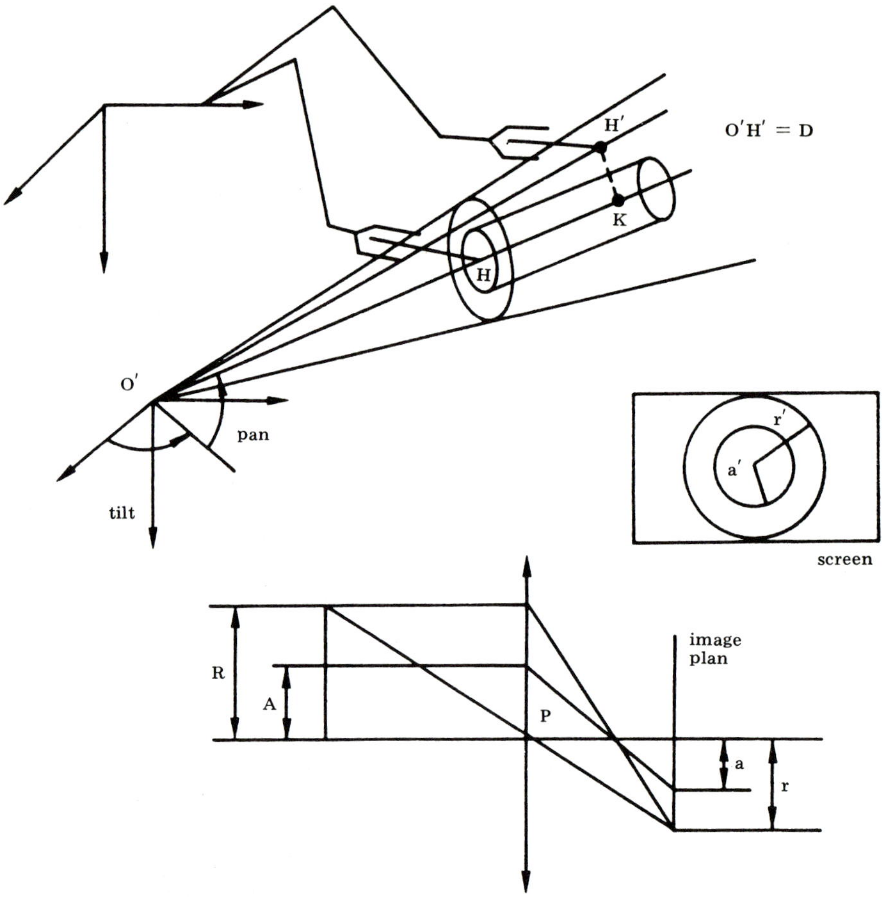

Figure 2.2.3.(7) *Conical field of vision of the camera*

N represents the reference surface in the plane of the camera image, that is, the surface that corresponds to the whole screen being covered, and it is therefore a constant. k is the variable parameter between 0 and 1, which allows realignment to be activated for a value of OH on the screen, greater than a certain distance. k can be displayed at the control station.

$N \geqslant 1$ is the condition for activating realignment.

Control of enlargement

This consists of applying at each instant the focal length which corresponds to enclosing, within the field of vision, a certain surface in a plane passing through the point of interest and at right angles to the optical axis of the camera. This surface can be displayed by the operator. It corresponds, therefore, to the choice of an enlargement to which the system must conform whatever the camera-object distance D, and the enlargement G is equivalent to $f/(D-f)$. Since G is imposed and D is measured, f can be adjusted so that, at each instant:

$$f = GD/(1+G) \qquad (2.2\text{-}16)$$

For the order of magnitude, if the camera-object distance varies between 0.3 and 3 m and the zoom can extend from $f = 10$ mm to $f = 60$ mm, then an enlargement of between 0.0034 and 0.25 can be displayed.

Automatic focusing

To regulate the focus, four properties must be taken into account:

1. camera-object distance D;
2. focal length f;
3. diameter of the circle of confusion c;
4. opening of the aperture n.

For a given focal length, the image is sharp at a distance of between:

$d_{min} = HD/(H+D)$, and
$d_{max} = HD/(H-D)$

where H is the hyperfocal length.
$H = f^2/cn$ (for $D = H$ the image appears sharp between $H/2$ and infinity).

If the camera has an automatic aperture, inaccessible to the operator, which is often the case, then calculations are made for the hypothetical worst conditions (maximal opening), using the following procedure:

1. calculation of D;
2. noting of f;
3. calculation of d_{min} and d_{max};
4. if $d_{min} < D < d_{max}$ no action is taken;
5. if $d_{min} \geqslant D \geqslant d_{max}$ the set of lenses is moved ad hoc by an amount calculated as a function of D and f.

Control strategy for the variables

The computer must, therefore, note the configuration of the slave arm, the state of the camera and of the control station at each instant, evaluate whether the orientation of the camera, the focusing and the zoom have the correct values and, if not, act to rectify them. The following can be displayed at the control station:

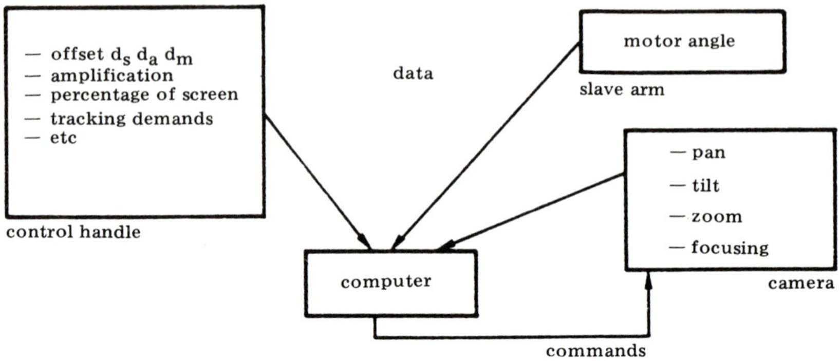

Figure 2.2.3.(8) *Camera positioning: general layout of the system*

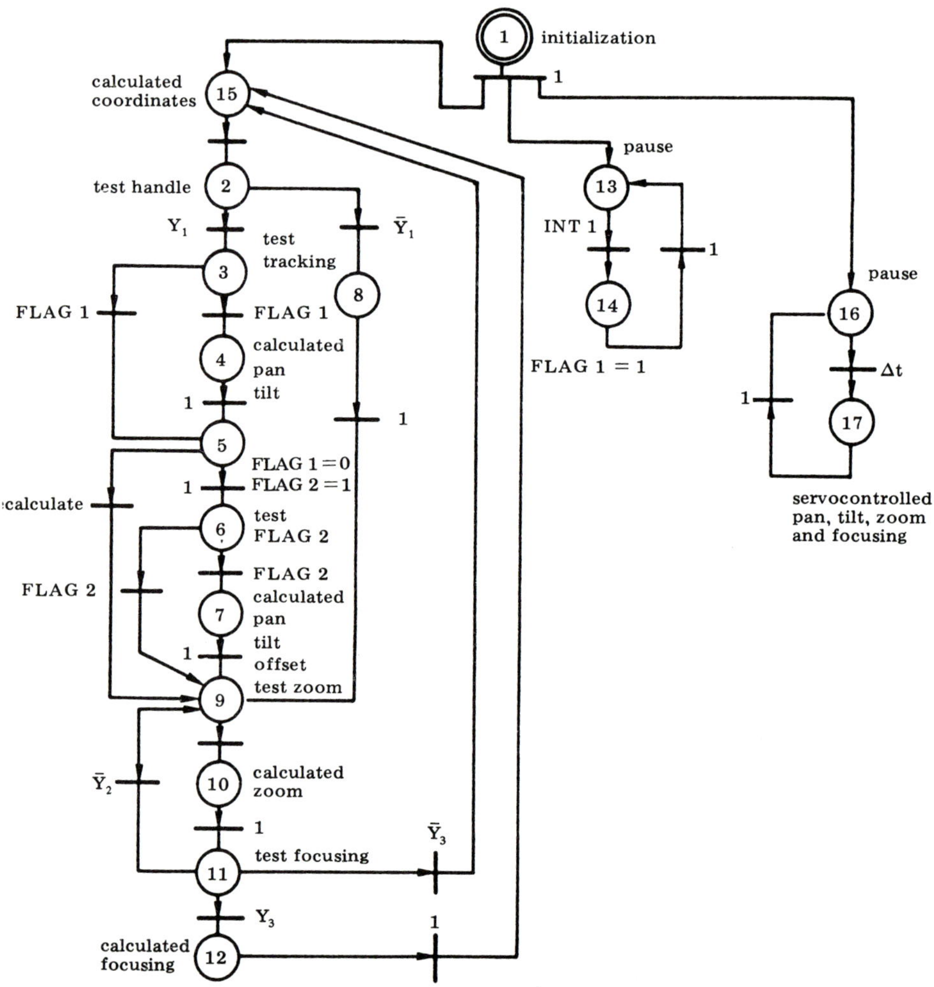

Figure 2.2.3.(9) *Grafcet*

1. shift of point H with respect to the centre of the gripper: (d_s, d_a, d_m);
2. enlargement G;
3. percentage of the screen defining automatic realignment.

Immediate camera positioning can also be requested [see Figure 2.2.3.(8)]. A command controller can be developed to deal with certain choices concerning:

1. parallelism of the calculations and commands;
2. order of commands in the motors concerned;
3. mode of control for the motors etc.

An example of this type of controller, integrated into a microprocessor, is shown in Figure 2.2.3.(9) in the form of Grafcet (Prunet, 1982).

Conclusions

The previous example was chosen to illustrate the fact that one of the important functions of the computer in teleoperation is coupling. This can be established between elements which are not coupled but would profit from being so. With two remote manipulation systems that are identical apart from the fact that one is equipped with a mobile camera controlled by the operator while the other has a camera of the type described above, the duration for execution of a task can be reduced by 40 per cent, with all other parameters kept the same.

2.2.4 Modification to the master-slave transmissions for movement or force (to help manipulation)

In a mechanical remote manipulation system, the position of the slave part is deduced from the corresponding part of the master, by a translation (sometimes geometrically proportional) followed by a rotation, that is a similitude.

If the master handle is pushed forwards, the slave gripper moves in the same direction etc. The amplification of movements depends on the application involved, but it is fixed for a given system (and generally work is carried out with a ratio of 1).

This homogeneity has a mechanical basis and the similarity of the movements is real for those with the same amplitude as the operator's movements.

This is not true for an MA11-type mechanical manipulator (see Section 1.3.4.3) in which transmission is not linear, and results in a spatial distortion.

The same phenomenon arises when the mechanical remote manipulator is motorized and bilaterally servocontrolled, by the simple addition of motors, retaining the master arm with identical structure to that of the slave arm.

With motorized unilateral manipulators, on the other hand, the master system structures are entirely different to those of the slave system (Cartesian or articulated, large transporter or manipulator) (see Section 1.4.5).

The master side generally includes:

1. potentiometers, each one acting on a different joint of the slave arm (for velocity or position);
2. push-buttons fulfilling the same role.

In certain models, the more pressure applied to the lever, the greater the velocity of displacement in the articulation concerned (velocity control) [see Figure 1.4.5.(2)]. Coordinated control takes the form of:

1. push-buttons, with each one corresponding to one of the following instructions to the gripper: move to the right, left, up, down, forward, back and rotate about each of the three axes of the gripper set of coordinate axes [see Figure 2.2.4.(1)].
2. control levers or spheres known as syntaxers (or joysticks), with three or six DOF, generally with a small angular or linear amplitude [see Figure 1.4.5.(8)]. The action of the operator on these handles can manifest itself in a displacement (positional or velocity control), or in a force or torque (velocity control) transmitted to the slave arm.

These electromechanical control modes do not necessarily imply the use of a digital computer, because the electronics implied is limited and fairly simple. (Of course, when a computer is used, these commands will generally be imbedded in it, but this is not an absolute necessity.) For the previous mechanical and motorized systems, the concept of 'normal' control can be qualified. This can be illustrated by an example to the effect that, if the 'forward movement of the gripper' button is

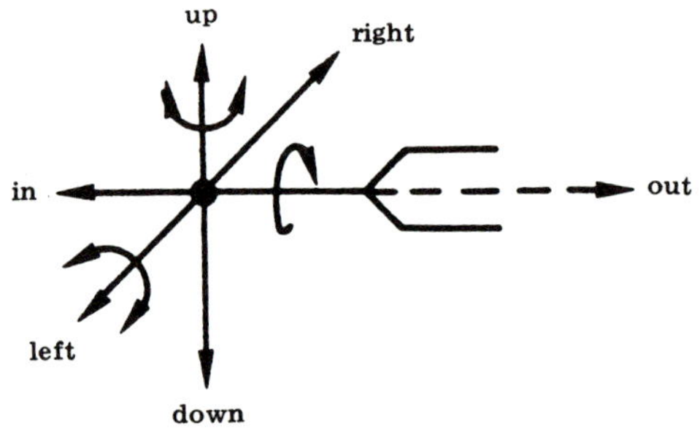

Figure 2.2.4.(1) *Coordinated control of the gripper using push-buttons*

depressed, this translation is always carried out and always in the same way (with the exception of cases in which an articular limit is encountered, and those in which the inherent lack of homogeneity in electromechanical system is involved when the set of accessible configurations is described).

The substitution of the computer for the human operator will be based on its capacity to alter the master-slave transmission as a function of aims or criteria which are better than the immediate aims or criteria of the human operator. The operator would often be incapable of taking these superior considerations into account when controlling the system.

This property of alteration due to the presence of the computer can present itself in a number of ways:

Case 1: the master part does not have the same number of DOF as the slave part. The operator may control, for example, one, two or three DOF while the computer controls the others according to the aims or criteria assigned to it.

Case 2: the master and slave arms are articulated with the same number of DOF. The operator controls the system as normal and the computer ensures:

case 2a: coordination of the DOF of the slave arm as distinct from that of the master arm;

case 2b: or a deviation in the trajectories when obstacles or the articular limits are being approached;

case 2c: or a feedback of artificial forces to the operator who feels the force to be applied for the execution of 'forbidden' gestures increase considerably, or on the other hand, decrease considerably for the execution of 'recommended' gestures (eg attraction to an object to be grasped);

case 2d: or modification of the behaviour of the tool.

Other, and perhaps more specific, cases can be envisaged. The ones given above are the most common.

These alterations to the transmission of the movements (and/or force) between master and slave may involve the use of sensors; this constitutes on-line alteration. They may also be based only on previously established data (eg to keep a tool in a vertical position); under these circumstances, there can be said to be an *a priori* alteration.

2.2.4.1 EXAMPLE OF CASE 1: A SYNTAXER WITH THREE DOF CONTROLLING A SLAVE ARM WITH SIX DOF

The reader should read this section in conjunction with Figure 1.4.5.(8) in Volume 3A of this series. This may be applied either in order to simplify the master structure, or to allow a handicapped person to

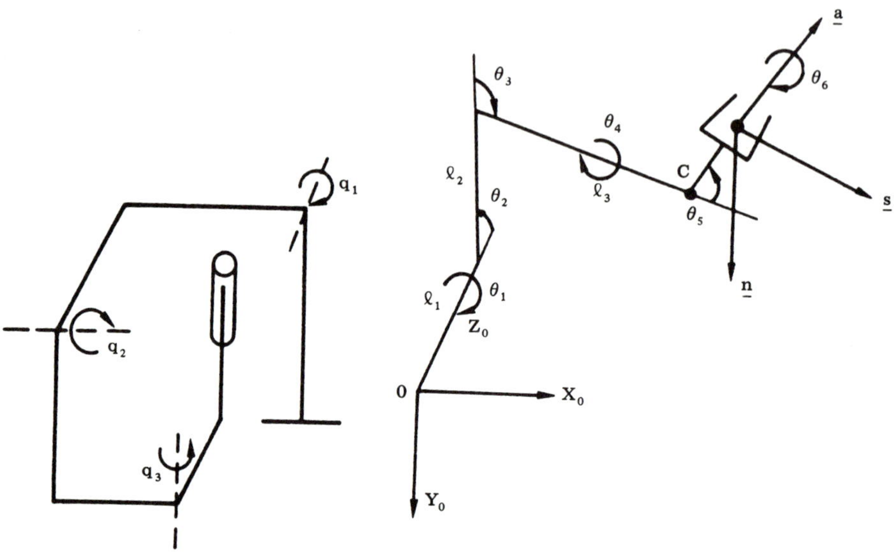

Figure 2.2.4.(2) *A syntaxer with three DOF controlling a slave arm with six DOF (from Liege, 1980)*

operate a manipulator unit by using his or her remaining motor functions. This is the case with the head-controlled telethesis produced during the Spartacus project (Guittet, 1985I; Guittet *et al.*, 1978, 1981).

The example under consideration concerns the control syntaxer for teaching a robot (Liege, 1980) [see Figure 2.2.4.(2)]. During this training period it is common to adopt a mode which constitutes, in practice, remote manipulation and, of course, the principle of this system is valid for remote manipulation as well (Guittet and Parent, 1978).

The syntaxer is made up of a handle with three rotation DOF (q_1, q_2, q_3) about the three normal axes. By using the computer it is possible to calibrate the actions applied to q_1, q_2 and q_3 on any three of the six DOF $\theta_1 \ldots \theta_6$ of the manipulator, giving 20 combinations and 120 possible arrangements. The arrangements and combinations of any use are somewhat smaller in number. The man-machine interface determines the choice. In practice, efforts are made to position the gripper (ie point C) using θ_1, θ_2 and θ_3, and to orient it using θ_4, θ_5 and θ_6. Therefore, q_1, q_2 and q_3 are immediately attributed to the variables θ_1, θ_2 and θ_3 or θ_4, θ_5 and θ_6. The three variables which are not linked to q_1, q_2 and q_3 are the responsibility of the computer which controls them as a function of:

1. the aim in view, or
2. the e olution of the three variables linked to q_1, q_2 and q_3.

The most important aspect from the operator's point of view (or at least the aspect that helps him considerably in piloting) is that of being able to manoeuvre with respect to the sets of coordinate axes that appear most natural to him. His main interest relates to two types of set: that on which he acts directly and that in which he sees the result of his action.

In the latter case, it is known that people are particularly sensitive to in-out, left-right and up-down movements. These movements constitute a Cartesian set of coordinate axes in which the up-down axis is vertical. For rotations, it appears that human operators can easily comprehend those that take place about the three normal axes of 'the object in rotation'. This is the case with the gripper set of coordinate axes a, n, s.

In relation to the set of coordinate axes on which he acts, the operator detects a certain degree of homogeneity between his own movement and that of the remotely operated system. In this example, this is possible when the rotations q_1, q_2 and q_3 give rise to rotation about the set of coordinate axes a, n, s.

This no longer applies when the rotations q_1, q_2 and q_3 bring about translations of C.

Finally, it should be noted that the operator is also sensitive to the uncoupling of control: acting on one DOF to obtain an up-down movement of C, and on another DOF for in-out movement etc, avoids the operator having too great a work load for the coordination of the DOF in the master system.

In short, the most important functions would appear to be the following:

1. Addressing q_1, q_2 and q_3:
   - to achieve displacement of C with the application of:
     q_1, translation along X_0;
     q_2, translation along Y_0;
     q_3, translation along Z_0;
   - to achieve orientation of the gripper:
     about the axes of the natural set of of coordinate axes a, n, s;
     about the other effective axes (eg those of a set parallel to $OX_0Y_0Z_0$, at the centre of the gripper.
2. Variation in the gain between a rotation of a q_i and the amplitude of the resulting movement. This gain may be fixed and displayable (as in the present example), or variable as a function of dynamic considerations of the way in which the syntaxer is handled (eg jerky movements). In the latter situation, the computer alters the control transmission (eg limitation of velocity, 'regularization' of the slave trajectory).
3. Coordination of the three DOF on which q_1, q_2 and q_3 do not act with the three DOF linked to q_1, q_2 and q_3, as a function of a displayed objective.

Liege (1980), for example, chose the following possibilities:

1. (q_1, q_2 and q_3) allowing C to be moved in parallel with respect to OX_0, OY_0, OZ_0, and the gripper maintaining a constant orientation in $OX_0Y_0Z_0$. (It moves parallel to itself.)
2. C remains fixed; q_1, q_2 and q_3 respectively in control about OX_0, OY_0, OZ_0.
3. C remains fixed; q_1, q_2 and q_3 respectively in control of rotations θ_4, θ_5 and θ_6.
4. q_1, q_2 and q_3 bring about the movement of C parallel to OX_0,OY_0, OZ_0. The gripper is moved passively by the mechanical structure.

Many other possibilities can be defined, and some of these will be given later, but for the time being, the translation of the various examples at the algorithmic level will be considered.

Displacement of C parallel to the axes $OX_0Y_0Z_0$

To establish an idea of the basic calculations performed by the computer, it should be noted that in the fixed set of coordinate axes $OX_0Y_0Z_0$, point C is defined by the coordinates:

$$X_c = S1S2\ell_2 + S1S(2+3)\ell_3 \quad (2.2\text{-}17)$$

$$Y_c = C1S2\ell_2 - C1S(2+3)\ell_3 \quad (2.2\text{-}18)$$

$$Z_c = \ell_1 + C2\ell_2 + C(2+3)\ell_3 \quad (2.2\text{-}19)$$

(see also Figure 2.2.4.(2) and the standard method given in Volume 1 of this series entitled *Modelling and Control*). So, by differentiating:

$$\Delta X_c = [C1S2\ell_2 + C1S(2+3)\ell_3]\Delta\theta_1 + [S1C2\ell_2 + S1C(2+3)\ell_3]\Delta\theta_2 + S1C(2+3)\ell_3\Delta\theta_1 \quad (2.2\text{-}20)$$

$$\Delta Y_c = [S1S2\ell_2 + S1S(2+3)\ell_3]\Delta\theta_1 + [C1C2\ell_2 + C1C(2+3)\ell_3]\Delta\theta_2 + [C1C(2+3)\ell_3]\Delta\theta_3 \quad (2.2\text{-}21)$$

$$\Delta Z_c = [S2\ell_2 + S(2+3)\ell_3]\Delta\theta_2 - S(2+3)\ell_3\Delta\theta_3 \quad (2.2\text{-}22)$$

to simplify the notation, let it be stated that:

$$\Delta X_c = \sum_{i=1}^{3} \alpha_i\Delta\theta_i; \quad \Delta Y_c = \sum_{i=1}^{3} \beta_i\Delta\theta_i; \quad \Delta Z_c = \sum_{i=1}^{3} \gamma_i\Delta\theta_i \quad (2.2\text{-}23)$$

Equation (2.2-23) can be inverted to show which $\Delta\theta_i$ will satisfy the ΔX_c, ΔY_c, ΔZ_c imposed. This gives:

$$\begin{pmatrix}\Delta\theta_1\\ \Delta\theta_2\\ \Delta\theta_3\end{pmatrix} = \frac{1}{\Delta}\begin{pmatrix}(\beta_2\gamma_3-\gamma_2\beta_3) & -(\alpha_2\gamma_3-\alpha_3\gamma_2) & (\alpha_2\beta_3-\alpha_3\beta_2)\\ -(\beta_1\gamma_3-\gamma_1\beta_3) & (\alpha_1\gamma_3-\alpha_3\gamma_1) & -(\alpha_1\beta_3-\alpha_3\beta_1)\\ (\beta_1\gamma_2-\gamma_2\beta_1) & -(\alpha_1\gamma_2-\alpha_2\gamma_1) & (\alpha_1\beta_2-\alpha_2\beta_1)\end{pmatrix}\begin{pmatrix}\Delta X_c\\ \Delta Y_c\\ \Delta Z_c\end{pmatrix}$$

(2.2-24)

$$\Delta = \alpha_1(\beta_2\gamma_3 - \gamma_2\beta_3) - \alpha_2(\beta_1\gamma_3 - \beta_3\gamma_1) + \alpha_3(\beta_1\gamma_2 - \beta_2\gamma_1) \qquad (2.2\text{-}25)$$

By bringing about a variation Δq_1 of q_1, a displacement is to be made parallel to OX_0, that is:

$$\Delta X_c = \ell_x, \quad \Delta Y_c = 0, \quad \Delta Z_c = 0$$

A variation Δq_1 will, therefore, bring about a variation of θ_1, θ_2 and θ_3 which should verify:

$$\Delta\theta_1 = (\beta_2\gamma_3 - \gamma_2\beta_3)\Delta\ell_x/\Delta \qquad (2.2\text{-}26)$$

$$\Delta\theta_2 = -(\beta_1\gamma_3 - \beta_3\gamma_1)\Delta\ell_x/\Delta \qquad (2.2\text{-}27)$$

$$\Delta\theta_3 = (\beta_1\gamma_2 - \beta_2\gamma_1)\Delta\ell_x/\Delta \qquad (2.2\text{-}28)$$

The gain will be characterized by the ratio $\Delta\ell_x/\Delta q_1$ which can be adjusted by means of the electronics and/or computer. In the same way, it is possible to produce a variation $\Delta Y = \Delta\ell_y$ by using a Δq_2, and a displacement $\Delta Z_c = \Delta\ell_z$ parallel to OZ_0 is acted on if a Δq_3 brings about:

$$\Delta\theta_1 = (\alpha_2\beta_3 - \alpha_3\beta_2)\Delta\ell_2/\Delta \qquad (2.2\text{-}29)$$

$$\Delta\theta_2 = -(\alpha_1\beta_3 - \alpha_3\beta_1)\Delta\ell_2/\Delta \qquad (2.2\text{-}30)$$

$$\Delta\theta_3 = (\alpha_1\beta_2 - \alpha_2\beta_1)\Delta\ell_2/\Delta \qquad (2.2\text{-}31)$$

It is important to note that α_i, β_i and γ_i are functions of θ_1, θ_2 and θ_3. They must be calculated on-line after taking values for θ_i at every instant. In addition, the determinant Δ is seen to appear in each of the expressions of the values $\Delta\theta_i$. If Δ equals zero, a singularity has arisen. This problem and the methods of dealing with it are discussed in Chapter 6, Volume 1 of this series entitled *Modelling and Control.*

Orientation of the gripper

This may be defined by several types of relationship: standard Euler angles (Landau and Lifchitz, 1960), modified Euler angles (Pradin, 1971), Euler parameters (Renaud, 1980), and components of the unitary vectors for the axes of the trihedron (a, n, s) along OX_0, OY_0 and OZ_0. The last of these representations, if redundant when the nine implied direction cosines are used, has the advantage of being relevant for a teleoperation system in which there is a tendency to refer to a Cartesian system with one vertical axis (in other words, the set of coordinate axes $OX_0Y_0Z_0$); for example, if the gripper is to be kept vertical and raised upwards, this condition is expressed immediately by:

$a_z = -1$ or $a_x = a_y = 0$;
s_x, s_y, s_z, n_x, n_y, n_z with any value.

By adopting this type of representation, the nine components implied in the manipulator shown in Figure 2.2.4.(2) can be expressed as follows, using the normal method:

$$a_x = C1S4S5 + S1C(2+3)C4S5 + S1S(2+3)C5 \quad (2.2\text{-}32)$$

$$a_y = S1S4S5 + C1C(2+3)C4S5 + C1S(2+3)C5 \quad (2.2\text{-}33)$$

$$a_z = +S(2+3)C4S5 + C(2+3)C5 \quad (2.2\text{-}34)$$

$$n_x = -C1C4S6 + C(2+3)S4S6 - C6[C1S4C5 + S1C(2+3)C4C5 + S1S(2+3)S5] \quad (2.2\text{-}35)$$

$$n_y = S6[S1C4 + C1C(2+3)S4] - S6[S1S4C5 - C1C(2+3)C4C5 + C1C(2+3)S5] \quad (2.2\text{-}36)$$

$$n_z = -S(2+3)S4S6 + C6[S(2+3)C4C5 + C(2+3)S5] \quad (2.2\text{-}37)$$

$$s_x = C6[C1C4 - S1C(2+3)S4] - S6[C1S4C5 + C(2+3)C4C5 - S1S(2+3)S5] \quad (2.2\text{-}38)$$

$$s_y = C6[S1C4 + C1C(2+3)S4] - S6[S1S4C5 + C1C(2+3)C4C5 + C1C(2+3)S5] \quad (2.2\text{-}39)$$

$$s_z = C6S(2+3)S4 + S6[S(2+3)C4C5 + C(2+3)S5] \quad (2.2\text{-}40)$$

These nine components are not independent, because of the existence of the following six relationships:

$$a_x^2 + a_y^2 + a_z^2 = 1 \quad (2.2\text{-}41)$$

$$n_x^2 + n_y^2 + n_z^2 = 1 \quad (2.2\text{-}42)$$

$$s_x^2 + s_y^2 + s_z^2 = 1 \quad (2.2\text{-}43)$$

$$a_x s_x + a_y s_y + a_z s_z = 0 \quad (2.2\text{-}44)$$

$$a_x n_x + a_y n_y + a_z n_z = 0 \quad (2.2\text{-}45)$$

$$s_x n_x + s_y n_y + s_z n_z = 0 \quad (2.2\text{-}46)$$

which indicate the perpendicularity of the vectors, taken two by two, and the unitary value of their norm.

Maintaining constant orientation of the gripper with respect to $OX_0Y_0Z_0$, whatever the position of C

This means that the gripper moves parallel to itself. The direction cosines must remain constant, and this is achieved simply by ensuring that three independent elements from among them are kept constant.

If, for example, a_x, a_z and n_z are chosen, their variation in any simple displacement must equal zero, so:

$$\begin{aligned}\Delta a_x = 0 = &[-S1S4S5 + C1C(2+3)C5]\Delta\theta_1 \\ &+ [-S1S(2+3)C4S5 + C(2+3)C5)][\Delta\theta_2 + \Delta\theta_3] \\ &+ [C1C4 - S1C(2+3)S4]S5\Delta\theta_4 \\ &+ [C1S4C5 + S1C(2+3)C4C5 - S15(2+3)S5]\Delta\theta_5\end{aligned} \quad (2.2\text{-}47)$$

$$\begin{aligned}\Delta a_z = 0 = &-[C(2+3)C4S5 + S(2+3)C5][\Delta\theta_2 + \Delta\theta_3] \\ &+ S(2+3)S4S5\Delta\theta_4 \\ &+ [S(2+3)C4C5 - S(2+3)S5]\Delta\theta_5\end{aligned} \quad (2.2\text{-}48)$$

$$\begin{aligned}\Delta n_z = &[-C(2+3)S4S6 + C(2+3)C4C5S6 - S(2+3)S5C6][\Delta\theta_2 + \Delta\theta_3] \\ &+ [-S(2+3)C4S6 - S(2+3)S4C5C6]\Delta\theta_4 \\ &+ [-S(2+3)C4S5C6 + C(2+3)C5C6]\Delta\theta_5 \\ &-[S(2+3)S4C6 + S(2+3)C4C5S6 + C(2+3)S5S6]\Delta\theta_6 = 0\end{aligned} \quad (2.2\text{-}49)$$

Equations (2.2-47) to (2.2-49) can be expressed in the following reduced form:

$$\Delta a_x = \sum_{i=1}^{5} \alpha_i \Delta\theta_i = 0 \quad (2.2\text{-}50)$$

$$\Delta a_z = \sum_{i=2}^{6} \beta_i \Delta\theta_i = 0 \quad (2.2\text{-}51)$$

$$\Delta n_z = \sum_{i=2}^{6} \gamma_i \Delta\theta_i = 0 \quad (2.2\text{-}52)$$

It has already been shown that the values q_i act on θ_1, θ_2 and θ_3 via equations (2.2-26) to (2.2-28) or (2.2-29) to (2.2-31). For each simple displacement, and in addition to the measured values of θ_1 to θ_6, the increments $\Delta\theta_1$, $\Delta\theta_2$ and $\Delta\theta_3$ are known. The desired variations of θ_4, θ_5 and θ_6 will, therefore, be deduced from equations (2.2-50) to (2.2-52), which are expressed:

$$\alpha_4\Delta\theta_4 + \alpha_5\Delta\theta_5 = -\alpha_1\Delta\theta_1 - \alpha_2(\Delta\theta_2 + \Delta\theta_3) = A \quad (2.2\text{-}53)$$

$$\beta_4\Delta\theta_4 + \beta_5\Delta\theta_5 = -\beta_2(\Delta\theta_2 + \Delta\theta_3) = B \quad (2.2\text{-}54)$$

$$\gamma_4\Delta\theta_4 + \gamma_5\Delta\theta_5 + \gamma_6\Delta\theta_6 = -\alpha_2\Delta\theta_2 - \alpha_3\Delta\theta_3 - \gamma_4\Delta\theta_4) = C \quad (2.2\text{-}55)$$

therefore:

$$\Delta\theta_4 = (A\beta_5\gamma_6 - B\alpha_5\gamma_6)/\Delta \quad (2.2\text{-}56)$$

$$\Delta\theta_5 = (-A\beta_4\gamma_6 + B\alpha_4\gamma_6)/\Delta \quad (2.2\text{-}57)$$

$$\begin{aligned}\Delta\theta_6 &= [A(\beta_4\gamma_5 - \gamma_4\beta_5) + B(\alpha_5\gamma_4 - \alpha_4\gamma_5) + C(\alpha_4\beta_5 - \alpha_5\beta_4)]/\Delta \\ \Delta &= (\alpha_4\beta_5 - \alpha_5\beta_4)\gamma_6\end{aligned} \quad (2.2\text{-}58)$$

During each simple control cycle, the computer measures the θ_1, θ_2 and $\Delta\theta_3$ values with respect to the previous cycle (unless it calculates these $\Delta\theta_1$, $\Delta\theta_2$ and $\Delta\theta_3$ values as shown in the preceding paragraph), then calculates $\Delta\theta_4$, $\Delta\theta_5$ and $\Delta\theta_6$ as shown above, and applies the suitable command in the circumstances. In this case also, the nullity of the determinant brings about a singularity which should be detected and corrected using the means already mentioned.

Rotation of the gripper about OX_0, OY_0 and OZ_0 as a result of action on q_1, q_2 and q_3 respectively

For the movement to be a rotation about OX_0, it must simply be ensured that, for example, a_x and n_x remain constant, that is $\Delta a_x = \Delta n_x = 0$. Let us assume that point C is fixed:

$$\Delta a_x = \alpha_4 \Delta\theta_4 + \alpha_5 \Delta\theta_5 = 0 \quad \text{(2.2-59)}$$

$$\begin{aligned}\Delta n_x = & \{C1[S4S6 - C4C5C6] + S1C(2+3)[C4S6 + S4C5C6]\} \Delta\theta_4 \\ & + \{C1S4S5S6 + S1C6[C(2+3)C4S5 + S(2+3)C5]\} \Delta\theta_5 + \\ & \Delta\theta_6 \{C6[-C1C4 + S1C(2+3)S4] + \\ & S6[C1S4C5 + S1C(2+3)C4C5 - S1S(2+3)S5]\}\end{aligned} \quad \text{(2.2-60)}$$

that is:

$$\Delta n_x = \delta_4 \Delta\theta_4 + \delta_5 \Delta\theta_5 + \delta_6 \Delta\theta_6 = 0 \quad \text{(2.2-61)}$$

equations (2.2-59) and (2.2-61), for example, lead to the following solution:

$$\Delta\theta_4 = \Delta\theta_4 \quad \text{(2.2-62)}$$

$$\Delta\theta_5 = (-\alpha_4/\alpha_5)\Delta\theta_4 \quad \text{(2.2-63)}$$

$$\Delta\theta_6 = [\delta_5\alpha_4 - \delta_4\alpha_5)\alpha_5]\Delta\theta_4 \quad \text{(2.2-64)}$$

All that needs to be done, therefore, is to link the suitable gain of the variable q_1 to the motor controlling θ_4 and to that the coordination of θ_5 and θ_6 is carried out by the computer, using equations (2.2-63) and (2.2-64) (if $\alpha_5 \neq 0$).

The same principle is used to obtain a rotation of the gripper about OY_0 by manoeuvring q_2, and a rotation about OZ_0 based on q_3.

Other functions

Using the same technique it is possible to obtain other functions at will, for example.

Displacement of C using q_1, q_2 and q_3 while ensuring that the gripper is directed towards a given point M [see Figure 2.2.4.(3)]: The straight line CM must always carry the vector $\underline{a}$. The cosine directrix of CM are:

$$u_x = (X_M - X_C)/d \quad (2.2\text{-}65)$$

$$u_y = (Y_M - Y_C)/d \quad (2.2\text{-}66)$$

$$u_z = (Z_M - Z_C)/d \quad (2.2\text{-}67)$$

$$d = [(X_M - X_C)^2 + (Y_M - Y_C)^2 + (Z_M - Z_C)^2]^{1/2} \quad (2.2\text{-}68)$$

It is important, for example, to maintain:

$$\begin{aligned} a_x &= (X_M - X_C)/d \\ a_z &= (Z_M - Z_C)/d \end{aligned} \quad (2.2\text{-}69)$$

But X_M and Z_M are known and imposed, X_C, Z_C and d are known at every instant from measurement of θ_1, θ_2 and θ_3 and calculations. a depends only on θ_4 and θ_5, and the two equations that must be solved take the following form:

For a_x:

$$a_1(\theta_1)S4S5 + a_2(\theta_1\theta_2\theta_3)C4S5 + a_3(\theta_1\theta_2\theta_3)C5 = u_x(\theta_1\theta_2\theta_3) \quad (2.2\text{-}70)$$

For a_z:

$$b_1(\theta_2\theta_3)C4S5 + b_2(\theta_2\theta_3)C5 = u_z(\theta_1\theta_2\theta_3) \quad (2.2\text{-}71)$$

These two equations have several determinations and it is simpler to proceed, as before, with increments by differentiating between equations (2.2-70) and (2.2-71), as shown below:

$$\sum_{i-1}^{5} \alpha_i \,.\, \Delta\theta_i = \sum_{i=1}^{3} \frac{\delta u_x}{\delta\theta_i} \,.\, \Delta\theta_i \quad (2.2\text{-}72)$$

$$\sum_{i=2}^{5} \beta_i \,.\, \Delta\theta_i = \sum_{i=1}^{3} \frac{\delta u_z}{\delta\theta_i} \,.\, \Delta\theta_i \quad (2.2\text{-}73)$$

which provide $\Delta\theta_4$ and $\Delta\theta_5$ as a function of the values θ_i and $\Delta\theta_1, \Delta\theta_2$ and $\Delta\theta_3$, measured or calculated elsewhere. Thus, when the syntaxer is moved, C moves according to a chosen mode (eg q_1, q_2 and q_3 giving rise to θ_1, θ_2 and θ_3 respectively, or else, as above, giving rise to movements parallel to OX_0, OY_0 and OZ_0 respectively). Using equations (2.2-72) and (2.2-73), the computer will evaluate the displacements $\Delta\theta_4$ and $\Delta\theta_5$, sending the commands ad hoc. In this situation, θ_6 is driven passively, with the direction of pointing being independent to it.

The same function as above, but in addition the plane of the gripper is maintained at right angles to the vertical plane passing through CM: To achieve this a constraint on θ_6 is introduced, and this can be explained as keeping $\underline{n}$ in a vertical plane, or $\underline{s}$ in a horizontal plane. In addition to equations (2.2-70) and (2.2-71) or (2.2-72) and (2.2-73), those corresponding to the following are present:

$$s_y = 0 \tag{2.2-74}$$

or

$$s_y = f_1(\theta_1\theta_2\theta_3\theta_4)C6 + f_2(\theta_1\theta_2\theta_3\theta_4\theta_5)S6 \tag{2.2-75}$$

$$\theta_6 = \text{arc tan}\,\frac{-f_1}{f_2} \tag{2.2-76}$$

which gives a (normal) determination to π, or thereabouts, easily raised if a process involves successive increments:

$$\Delta s_y = 0 = \sum_{i=1}^{5} g_i(\theta_1 \ldots \theta_6)\Delta\theta_i + g_6(\theta_1 \ldots \theta_6)\Delta\theta_6 \tag{2.2-77}$$

$$\Delta\theta_6 = -\sum_{i=1}^{5} g_i(\theta_1 \ldots \theta_6)\Delta\theta_i / g_6(\theta_1 \ldots \theta_6) \tag{2.2-78}$$

The same pitfalls will always be lined with any singularities for calculation and control when $g_6 \simeq 0$.

Displacement of C using q_1, q_2 and q_3 while ensuring that the gripper is horizontal: In this situation:

$$n_y = 1 \quad \text{or} \quad n_x = n_z = 0 \tag{2.2-79}$$

The same principle still applies for calculation. With an incremental procedure, it is necessary simply to verify the following equation after an initialization ensuring this horizontally (eg $\theta_1 = \theta_4 = \theta_6 = 0$) [see Figure 2.2.4.(3) in which the manipulator has the same definition as that given in Figure 2.8.4.(2)]:

$$\Delta n_y = 0$$

or

$$\sum_{i=1}^{6} h_i(\theta_1 \ldots \theta_6)\Delta\theta_i = 0 \tag{2.2-80}$$

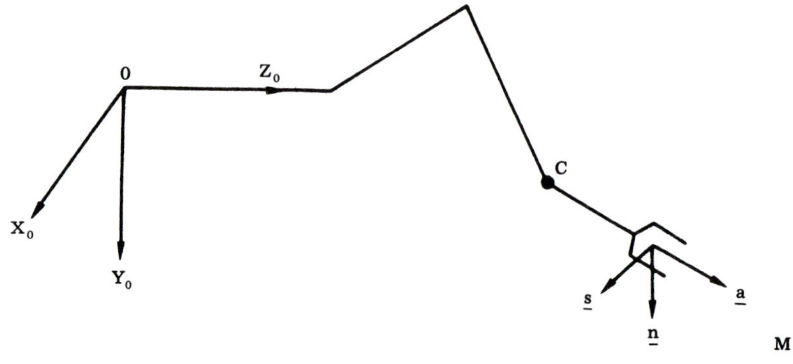

Figure 2.2.4.(3) *Displacement of C using q_1, q_2 and q_3 while ensuring that the gripper is horizontal*

Since action is carried out directly on θ_1, θ_2 and θ_3, equation (2.2-80) with three unknowns, $\Delta\theta_4$, $\Delta\theta_5$ and $\Delta\theta_6$, must be respected. The simplest choice, bearing in mind the mechanical structure under consideration, is:

$$\Delta\theta_4 = \Delta\theta_5 = 0 \tag{2.2-81}$$

$$\Delta\theta_6 = -\sum_{i=1}^{3} h_i(\theta_1 \ldots \theta_6)\Delta\theta_i / h_6(\theta_1 \ldots \theta_6) \tag{2.2-82}$$

Displacement of C with q_1, q_2 and q_3 on a sphere of centre M, with the gripper directed towards M and remaining at right angles to a vertical plane passing through CM: To the constraint equations (2.2-72), (2.2-73) and (2.2-78) must be added a constraint on θ_1, θ_2 and θ_3, expressed by:

$$|CM| = \text{constant or } \Delta|CM| = 0 \tag{2.2-83}$$

therefore:

$$\Delta[(X_M - X_C)^2 + (Y_M - Z_C)^2] = 0 \tag{2.2-84}$$

$$(X_M - X_C)\Delta X_C + (Y_M - Y_C)\Delta Y_C + (Z_M - Z_C)\Delta Z_C = 0 \tag{2.2-85}$$

By using equations (2.2-17) to (2.2-22), equation (2.2-85) can easily be written in the following form:

$$\sum_{i=1}^{3} \ell_i(\theta_1\theta_2\theta_3).\Delta\theta_i = 0 \tag{2.2-86}$$

If it is decided to link q_1 to θ_1, q_2 to θ_2 and q_3 to θ_3, then by acting on q_1 a $\Delta\theta_1$ is produced manually. The computer must act on the other two DOF to allow verification of equation (2.2-86), so:

$$\ell_2\Delta\theta_2 + \ell_3\Delta\theta_3 = -\ell_1\Delta\theta_1 \tag{2.2-87}$$

It is possible, for example, to choose:

$$\Delta\theta_2 = 0 \tag{2.2-88}$$

$$\Delta\theta_3 = -(\ell_1/\ell_3)\Delta\theta_1 (\ell_3 \neq 0) \tag{2.2-89}$$

In the same way, by acting on q_2, the following is brought about:

$$\Delta\theta_3 = 0 \tag{2.2-90}$$

$$\Delta\theta_1 = -(\ell_2/\ell_1)\Delta\theta_2 (\ell_2 \neq 0) \tag{2.2-91}$$

And finally, when q_3 is manoeuvred, the computer establishes the following state:

$$\Delta\theta_1 = 0 \tag{2.2-92}$$

$$\Delta\theta_2 = -(\ell_3/\ell_2)\Delta\theta_3 (\ell_2 \neq 0) \tag{2.2-93}$$

Note also the possibility of singularities arising if ℓ_1, ℓ_2 or ℓ_3 equal zero (see Volume 1 of this series entitled *Modelling and Control* for the appropriate solution). Equations (2.2-72), (2.2-73) and (2.2-78) indicate the behaviour to be followed by θ_4, θ_5 and θ_6.

This function may be useful in situations where a camera is linked to the handle of the manipulator, and is to be used to study an object, the position of which is known, from various angles.

Conclusions

The computer authorizes the programming of multiple alteration functions for movement transmission which are, in fact, also automatic coordination functions for the articular movements and useful to the operator in the performance of various tasks. The aspects that concern only the movements of the manipulator arm can easily be generalized as the coordination of the large transporter or the vehicle and, more importantly, as the processing of redundancies existing between this transporter and the manipulator.

2.2.4.2 EXAMPLES FOR CASE 2: ALTERATION IN 'NORMAL' COORDINATION BETWEEN THE MASTER AND SLAVE ARMS

Case 2a

The role of the computer in the previous system, where the master part had less DOF than the slave part, can be transposed to a situation in which both master and slave have the same number of DOF.

Using this hypothesis, uncoupling would take place between certain

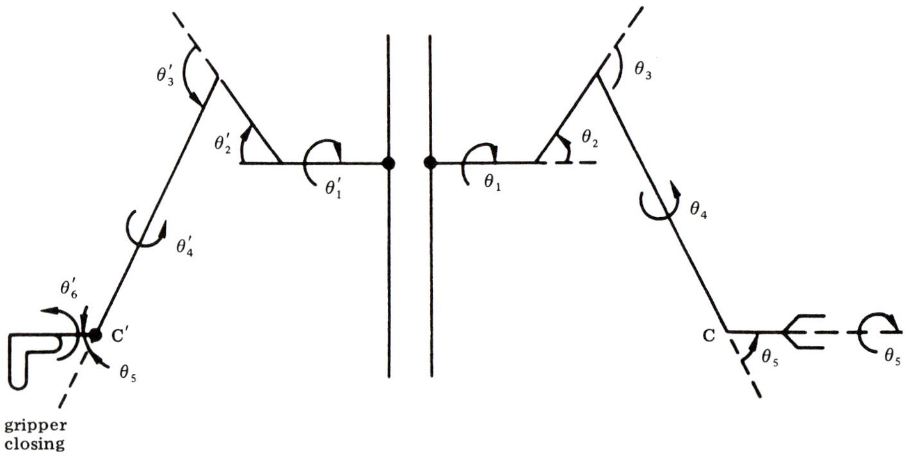

Figure 2.2.4.(4) *Standard master-slave system for case 2a*

DOF in the master and the corresponding DOF in the slave, with the computer automatically taking charge of these.

Consider the example of the standard system shown in Figure 2.2.4.(4). The normal correspondence is $\theta_i' \longrightarrow \theta_i$ (i = 1.6), but it is possible to alter the transmission by uncoupling either the first three DOF (which control the position of C) or the last three DOF (which control the orientation of the gripper) on demand, and replacing control of the uncoupled DOF with computer-based control fulfilling, for example, one of the functions mentioned in Section 2.2.4.1 [see Figure 2.2.4.(5)].

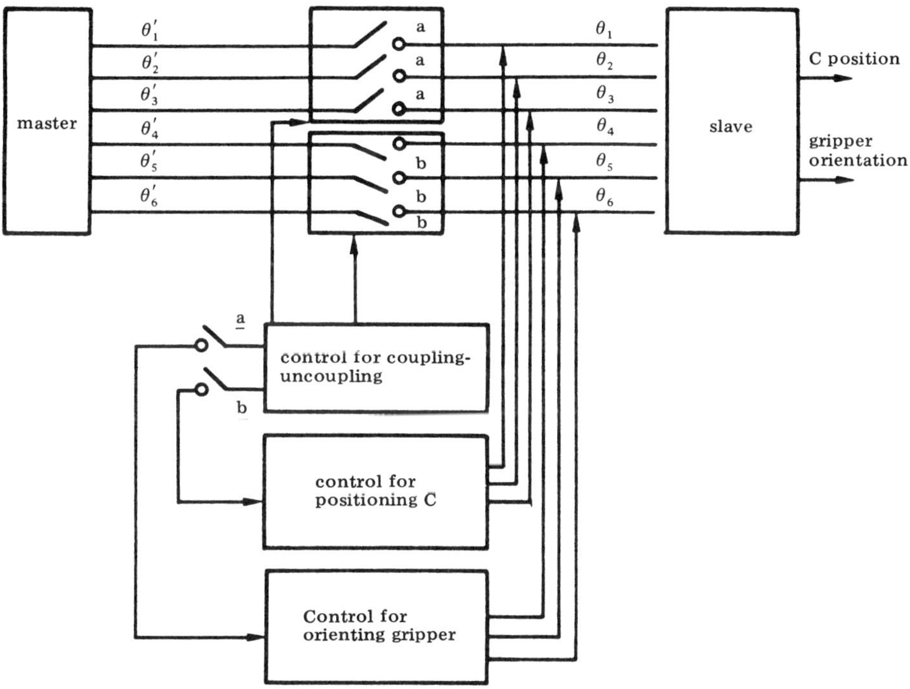

Figure 2.2.4.(5) *Alteration of transmissions using a computer*

Case 2b

The uncoupling mentioned above is carried out at the request of the operator who 'calls' the corresponding function when he judges it to be appropriate. A more sophisticated mode of operation may be achieved with automatic uncoupling in dangerous situations, with automatic selection of the appropriate function.

Take, for example, automatic avoidance of obstacles of which the computer had no advance knowledge (which implies the use of proximity sensors on the slave arm gripper).

In Figure 2.2.4.(6), a simple two-dimensional example (can easily be extended to three dimensions) is shown. Proximity sensors are situated on the extreme end of the gripper (Espiau, 1981; Espiau and André, 1981; Espiau and Catros, 1980) and allow the distance between each side of the gripper and any possible obstacle to be found (although not always very accurately), in two normal directions $\underline{a}$ and $\underline{s}$. When one or more distances are smaller than a predetermined threshold (the safety margin), uncoupling is activated with a displacement strategy. The DOF

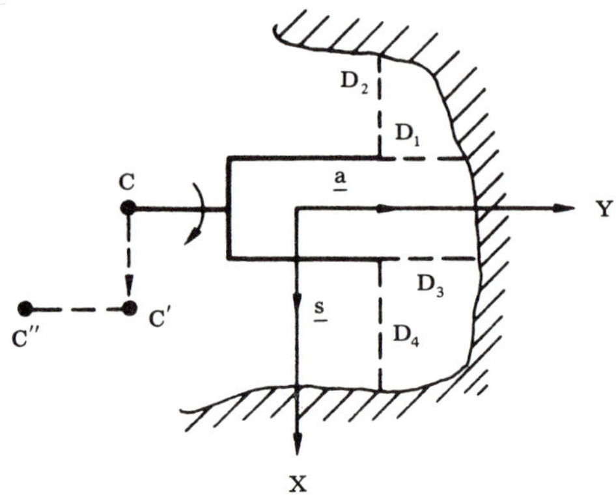

D_1	0	1	0	1	0	1	0	1	0	1	0	1	0	1	0	1
D_2	0	0	1	1	0	0	1	1	0	0	1	1	0	0	1	1
D_3	0	0	0	0	1	1	1	1	0	0	0	0	1	1	1	1
D_4	0	0	0	0	0	0	0	0	1	1	1	1	1	1	1	1
probable situations																
possible strategies	nothing	ΔX	ΔX	ΔX	$-\Delta X$	$\Delta\theta_3\pi/2$	$-\Delta Y + \Delta\acute{X} + \pi/4$	$-\Delta Y + \Delta X$	$-\Delta X$	$-\Delta Y - \Delta X - \pi/4$	$\Delta Y \Delta\theta$ enfdes Di	$-\Delta Y$	$-\Delta X$	$-\Delta X - \Delta Y - \pi/4$	$-\Delta Y$	$-\Delta Y$
case	1	2	3	4	5	6	7	8	9	10	11	12	13	14	15	16

Figure 2.2.4.(6) *Two-dimensional example of case 2b. The gripper is equipped with four proximity sensors. If D_i is smaller than the safe distance, 1 is selected for uncoupling. In the reverse situation 0 is selected*

implied in this strategy can no longer be controlled by the operator during this displacement.

There are many possible strategies, and suitable ones for each case in Figure 2.2.4.(6) have been chosen arbitrarily here. It should be noted that, in certain cases (2, 3, 4, 5, 8, 9, 12, 13, 15 and 16), the automatic displacement occurs parallel to either X or Y (bringing about the uncoupling of the DOF implied in the position of C). In case 6, the uncoupling implies the DOF controlling the orientation of the gripper. Cases 7, 10 and 11 show total uncoupling of the master and the slave.

The choices made here are only approximate, and far more precise strategies can be applied by not limiting the role of the sensors to the extent of making them act only as switches. If the real distance between the sensors and obstacles can be measured, very high-level strategic function can clearly be developed according to the needs of each designer (Espiau, 1981; André, 1984).

Consider, once again, Figure 2.2.4.(6) in case 8 for example. If the remote manipulation system shown in Figure 2.2.4.(4) is used, for a situation in which D_1, D_2 and D_3 are less than a previously determined limit distance, the strategy chosen will consist of uncoupling θ_1, θ_2 and θ_3 from θ'_1, θ'_2 and θ'_3, and having control of them taken over by the computer, which should bring about at point C the displacement ΔX (eg equal to the length of the gripper) and ΔY, both of which will be imposed. If these displacements are carried out one after the other, CC′ will take place parallel to $\underline{s}$, and the following equations can be established:

$$\Delta X_c = \Delta X . s_x = \sum_{i=1}^{3} a_i(\theta_1 \theta_2 \theta_3)\Delta\theta_i \qquad (2.2\text{-}94)$$

$$\Delta Y_c = \Delta X . s_y = \sum_{i=1}^{3} b_i(\theta_1 \theta_2 \theta_3)\Delta\theta_i \qquad (2.2\text{-}95)$$

$$\Delta Z_c = \Delta X . s_z = \sum_{i=1}^{3} c_i(\theta_1 \theta_2 \theta_3)\Delta\theta_i \qquad (2.2\text{-}96)$$

By referring to equations (2.2-20) to (2.2-22) and (2.2-41) to (2.2-43), and knowing that the θ_i values are measured at the time of uncoupling, equations (2.2-94) to (2.2-96) can be used to provide the values for $\Delta\theta_1$, $\Delta\theta_2$ and $\Delta\theta_3$. In the same way, the displacement C′C″ is made parallel to $\underline{a}$.

$$\begin{aligned} \Delta X_{C'} &= -\Delta Y a_x \\ \Delta Y_{c'} &= -\Delta Y a_y \\ \Delta Z_{c'} &= -\Delta Y a_z \end{aligned} \qquad (2.2\text{-}97)$$

are resolved as before, using equations (2.2-20) to (2.2-22) and (2.2-35) to (2.2-37).

There are three practical problems involved here:

1. The operator must be able to cancel the automatic uncoupling if he wishes to override the obstacle avoidance function. This situation might arise when these obstacles become objects of interest (eg objects to be grasped). This question will be considered later.
2. The operator should not have a feeling of helplessness at the time uncoupling takes place and the slave starts to move automatically. This is a complex problem involving the rapidity of intervention by the computer and also the physical and psychological sensations of the operator. Ideally, the operator should feel that he is cooperating with the computer rather than handing over to it. For this to be the case, the computer must, in fact, carry out what the operator 'would have liked to do' almost at the same time as he would have done it. So far there does not seem to have been any study made of this aspect, since most computer assistance has been developed in a context of exclusive intervention by one or other of the two decision making centres present.
3. After total or partial uncoupling between master and slave, and once the avoidance procedure has been completed, automatic recoupling takes place. During the uncoupled phase, the master DOF are controlled by the action of the operator. It is, therefore, not possible to establish electromechanical master-slave coupling of position-position type, but only of relative type. This means that θ_i' no longer corresponds to θ_i but that the relationship is based on the increments $\Delta\theta_i' - \Delta\theta_i$. If the real error between master and slave becomes too great, there is a risk that the problem of the articular limits, that is the possible range of movement for each articulation, may become increasingly important. As a result of this, it must be possible to reset the servocontrol relationship between the master and slave positions periodically, at the request of the operator (reinitialize the correspondence between $\underline{\Theta}$ and $\underline{\Theta}'$).

Case 2c: feedback of artificial forces

Another way in which the computer can assist the operator consists not of interrupting the master-slave transmission in appropriate circumstances, but of making proscribed displacements difficult for the operator to perform, and the recommended displacements very easy. This means that the forces that the operator must develop to execute a given manoeuvre are artificially increased or decreased as a function of the risk or advantage presented by the movement.

To achieve this, the computer must know which movements are proscribed and which are to be encouraged. There are three main ways of giving the computer this information:

1. *a priori* information, that is, the computer knows in advance the articular limits, the position and structure of obstacles, and the position of objects of interest etc;
2. autonomous on-line information provided by appropriate sensors independent of the operator;
3. on-line information provided by the operator who, by activating the computer functions, indicates to it, for example, that an object towards which he is moving is not an obstacle but an object of interest; the same (or other) sensors that act to bring about a strategy to help in object avoidance will then be used to develop a strategy to help in approach or grasping.

This presupposes an important property for the teleoperation system: that it should be completely bilateral, that is, the master system is motorized in such a way that the force can be fed back to its articulations.

Two different types of force can be distinguished, according to their interpretation:

1. Those corresponding to feedback to the master articulations: the real force felt on the slave side, either with the same intensity, or with a scale factor adapted to the operator (eg amplification if the slave system is a micromanipulator, and reduction if the system is being used to handle heavy loads). In electric teleoperation this can be done without the help of data processing. This mode of force feedback will be referred to as normal in any bilateral servo-system, as above [see Section 1.5.4] (Vertut, 1975, 1976, 1977; Vertut and Coiffet, 1977; Vertut *et al.*, 1976I, II).
2. Those corresponding to an alteration in this 'normal' force feedback, for reasons of operator assistance, as mentioned above: this alteration constitutes a form of aid which always involves the computer.

Figure 2.2.4.(7) shows the schema of a bilateral system. In the normal mode, the system is completely symmetrical (the scale factor, which may vary somewhat, is taken here as being equal to one) such that, if the slave is manoeuvred, the master is servoed to its position, and vice versa. If this servocontrol functions simultaneously for position and torque (shown in this example using electric torque-motors, by the currents I and I′), the following is found:

$$\underline{V} = \underline{f}(\underline{\Psi}', \underline{I}') \quad \text{and} \quad \underline{V}' = \underline{g}(\underline{\Psi}, \underline{I}) \qquad (2.2\text{-}98)$$

The functions f and g differ only by the scale factor and the mechanical couplings differentiating the master and slave structures.

In particular, if a force originating from outside the manipulator is applied to the slave gripper, but $\underline{\Psi}$ is kept constant (by the imposed $\underline{V}$),

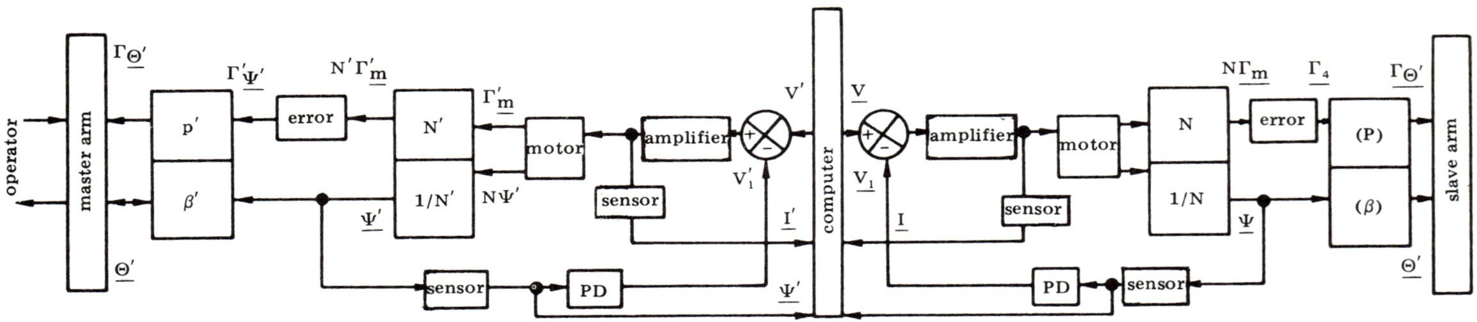

Figure 2.2.4.(7) *Schema of a bilateral servo manipulator system controlled by computer*

$\underline{\Gamma_\Theta}, \underline{\Gamma'_\Theta}$: *torque vectors for the slave and master joints;*

(P), (P′): *matrices with respect to* $\underline{\Gamma_\Psi}$ *and* $\underline{\Gamma'_\Psi}$, *and* $\underline{\Gamma_\Theta}$ *and* $\underline{\Gamma'_{\Theta'}}$;

(β), (β′): *matrices for mechanical coupling between* $\underline{\Psi}$ *and* $\underline{\Theta}$, *and* ψ' *and* θ';

$\underline{\Gamma_\Psi}, \underline{\Gamma'_{\Psi'}}$: *motor torques after motor friction, inertia etc have been considered;*

$\underline{\Gamma_m}, \underline{\Gamma'_m}$: *theoretical torque vectors on the slave and the master;*

N, N′: *demultiplication matrices between* ψ *and* $\underline{\Theta}$, *and* $\underline{\Psi'}$ *and* θ';

$\underline{I}, \underline{I'}$: *motor current vectors (torque varies with current:* $\underline{\Gamma_m} = (K)\underline{I}$; $\underline{\Gamma'_m} = (K')\underline{I'}$)

PD: *proportional and derivative compensation network*

the motor torques as well as the currents $\underline{I}$ will increase, and $\underline{V}'$ will also increase, which the operator will feel. This property is extremely useful in teleoperation and is currently found in numerous electrical systems, referred to as *force reflecting systems* or as having *force feedback.*

The role of the computer in the functions considered here will now consists of superimposing an artificial force feedback on the normal force feedback from the slave to the operator. Equation (2.2-102) will be replaced by:

$$\begin{aligned} \underline{V} &= \underline{f}(\underline{\Psi}', \underline{I}') \\ \underline{V}' &= \underline{g}[\underline{\Psi}, \underline{I}(\text{strategy})] \end{aligned} \qquad \textbf{(2.2-99)}$$

In other words, it is as a function of the considerations on the eventual objective and the configuration in which the system is finally placed that $\underline{I}$ and/or $\underline{V}'$ will vary artificially.

In this way, the operator can be assisted by opposition being presented to the performance of proscribed movements. Consider once again the example of the gripper equipped with four proximity sensors [see Figure 2.2.4.(6)]. If the gripper is moved to the same position as that shown in case 10, the proscribed movement will be:

ΔY positive, ΔX positive, $\Delta \theta$ positive.

These can be recognized by interpreting the signals from the proximity sensors. The action of the computer will then be such that the more the operator tries to displace the gripper along positive Y, the greater the force to be exerted from the master side. This case can be examined more closely.

Moving along positive Y amounts to satisfying relationships of a similar type to equation (2.2-97), that is:

$$k\, a_x(\theta_1 \ldots \theta_5) = \Delta X_C = \sum_{i=1}^{3} f^1(\theta_1\theta_2\theta_3).\Delta\theta_i \qquad \textbf{(2.2-100)}$$

$$k\, a_y(\theta_1 \ldots \theta_5) = \Delta Y_C = \sum_{i=1}^{3} f^2(\theta_1\theta_2\theta_3).\Delta\theta_i \qquad \textbf{(2.2-101)}$$

$$k\, a_z(\theta_2 \ldots \theta_5) = \Delta Z_C = \sum_{i=1}^{3} f^3(\theta_1\theta_2\theta_3).\Delta\theta_i \qquad \textbf{(2.2-102)}$$

which gives the values for $\Delta\theta_1$, $\Delta\theta_2$ and $\Delta\theta_3$ as:

$$\begin{aligned} \Delta\theta_i &= k\, g_i(\theta_1 \ldots \theta_5) \\ i &= 1,3 \end{aligned} \qquad \textbf{(2.2-103)}$$

The ratios between the three $\Delta\theta_i$ values are important, because they characterize the direction adopted by the gripper. If these ratios verify equations (2.2-100) to (2.2-102), the movement must be opposed, but it is clear that similar movements cannot be tolerated in the same way.

In practice, the whole zone defined by $\Delta Y > 0$, $\Delta X > 0$ and $\Delta Z > 0$ should be avoided, with movements defined according to equations (2.2-20) and (2.2-21) by:

$$\begin{aligned} &C1[S2\ell_2 + S(2+3)\ell_3]\Delta\theta_1 + S1[C2\ell_2 + C(2+3)\ell_3]\Delta\theta_2 \\ &\quad + S1C(2+3)\ell_3\Delta\theta_3 > 0 \end{aligned} \qquad \textbf{(2.2-104)}$$

$$\begin{aligned} &S1[S2\ell_2 + S(2+3)\ell_3]\Delta\theta_1 + C1[C2\ell_2 - C(2+3)\ell_3]\Delta\theta_2 \\ &\quad - C1C(2+3)\ell_3\Delta\theta_3 > 0 \end{aligned} \qquad \textbf{(2.2-105)}$$

and, assuming that rotation θ is taken care of by θ_4:

$$\Delta\theta_4 > 0 \qquad \textbf{(2.2-106)}$$

The computer may, therefore, record the various elements of these equations at each moment, and increase the part of $\underline{V}'$ applied to θ_1, θ_2, θ_3 and θ_4 exponentially, relative to the degree of overshooting. It does not modify $\underline{V}$ if none of equations (2.2-104) to (2.2-106) is verified. This method is particularly applicable if any obstacles are described before hand, and no sensors are used.

In the example under consideration, the sensors measure the distances, and there is a method for use in this situation involving perhaps slightly less calculation.

In case 10 [see Figure 2.2.4.(6)] distances D_1 and D_4 have become too small. The solution consists simply of preventing any movement which makes them smaller. Let D_i be the distance between a sensor and an obstacle. If a small movement is made with the master arm, let the distance now be D_i'.

If $D_i' \geqslant D_i$ no action is taken.

If $D_i' < D_i$ a supplementary element in the form $1/D_i'^n$ will be returned on the entire vector $\underline{V}'$. As D_i' decreases, the force required to reduce it becomes more intense from the operator's point of view. The force felt is regulated with n. By letting the arm move in the directions requiring the least energy, obstacles can easily be avoided.

The method can be made symmetrical if an object of interest is to be approached. This object will attract in proportion, for example, to D_i^2 by freeing $\underline{V}'$. The combination of the two principles makes it difficult to move away from the object of interest and easy to move towards it.

Using the same principle, the grasping of an object can be made easier, as shown by Espiau (1981; André, 1984), by using the difference in the sensor signals giving D_1 and D_3 (which allows the gripper to be centred relative to the object) and the signal from another sensor situated at the bottom of the gripper (making it easier to position the object within the gripper).

These principles can also be applied to avoid each articulation going right to its limit, either on the master or slave side. In this case, the fact that the obstacles are known in advance is an advantage. In addition, each limit implies only one θ_i when the articulations are mechanically uncoupled. It is not rare, however, to find that the ranges of certain articulations depend on the configuration of the manipulator.

For this example, the slave system shown in Figure 2.2.4.(4) is assumed to have the following ranges of movement:

$$\theta_{imin} \leqslant \theta_i \leqslant \theta_{imax} \tag{2.2-107}$$

for i = 1 to 3

$$\theta_{jmin}(\theta_1\theta_2\theta_3) \leqslant \theta_j \leqslant \theta_{jmax}(\theta_1\theta_2\theta_3) \tag{2.2-108}$$

for j = 4 to 6

An angular distance is fixed for each limit, from which action will be taken:

$$\Delta\theta_{im}, \Delta\theta_{jm}$$

At each moment the computer can evaluate:

$$|\theta_i - \theta_{imin}|, |\theta_i - \theta_{imax}|$$

for i = 1 to 3

$$|\theta_j - \theta_{jmin}(\theta_1, \theta_2, \theta_3)|, |\theta_j - \theta_{jmax}(\theta_1, \theta_2, \theta_3)|$$

for j = 4 to 6

If one of these values is less than $\Delta\theta_{im}$ ($\Delta\theta_{jm}$), an artificial force is applied to the corresponding motor on the master side, proportional, for example, to $1/|\theta_i - \theta_{imin}|^n$.

This provides a simple method of preventing the operator from forcing the slave arm to its limits (whether he is unaware or fails to notice that he is doing so) at the risk of damaging the system.

Case 2d: alteration in the behaviour of the tool

The aim of teleoperation is to use the master-slave system to move a tool into the desired position and orientation with respect to the environment, and to make it work. If the tool in question is a gripper, its work will consist of grasping an object between its jaws; if it is a buff wheel, it may be used to rub down a component etc.

In all these types of work, the control of the tool cannot always be carried out in the on-off mode. The interaction between the terminal device and the environment must often be checked, and the computer can act as the intermediary between the operator and the tool to correct or complete the control action of the operator to ensure the efficient execution of the task in hand.

Consider the case of a gripper used as a tool. The closing of the jaws is often controlled by the operator using a trigger to provide a certain degree of control over the grasping force. This is only approximate, however, and in some cases closing and opening are controlled simply by a switch.

In this situation the computer can also provide assistance by adjusting the grasping force to the weight of the object being carried. The gripper shown in Figure 2.2.2.(2) is equipped with sensors that give it this property. By detecting if the object slips, the grasping force can be modified in such a way as to regulate it at the lowest level compatible with the external forces (such as gravity) (Stojilkovic and Clot, 1977; Peruchon, 1979).

Conclusions

This chapter is intended to show at least part of the extensive aid that can be provided by the use of a second brain, the computer, to the first brain, which is that of the human operator in teleoperation. This contribution differs from that used in controlling robots (although often having recourse to the same types of technique), in that it is based specifically on cooperation with the operator or correction of actions taken by the operator rather than on substitution pure and simple (which only occurs for short periods during execution of the task).

To a certain extent, the computer contributes artificially to the operator's reflexes, and takes over certain functions in parallel with him, thus reducing the level of physical and mental fatigue experienced by the operator.

The computer can, however, do more than understand the actions of the operator and interpret them to make the execution of the task easier. It can also help the operator by keeping him permanently informed of important facts relating to the task. This other contribution forms the subject of Chapter 2.3.

Chapter 2.3

The use of computer feedback to the operator

2.3.1 Introduction

The aim is to use the computer to place easily understood and pertinent data relating to the conditions for the execution of the task at the disposal of the operator.

Almost every word of this definition constitutes a problem in itself. To date, all efforts in this field have simply been heuristic, sometimes with successful results. At the theoretical level, a tremendous amount has still to be achieved (although this situation is not unusual, since in fields of research involving such complex terms as operator or task, theory generally derives from experience and not from reasoning alone, since theoretical results cannot be verified).

This chapter is intended to throw some light on a number of important questions raised by this subject.

2.3.2 Making use of the senses

The human sensory system is known to be characterized by its immense diversity (Loewenstein, 1971), even though in common parlance it is reduced to five main functions. During execution of a manual task, it can be said that the senses of sight and touch are generally used over and above any of the others. In reality, all the senses are used but sight and touch can be considered as the active functions allowing execution of the task to take place, and the other senses can be considered as passive functions taking the role of watch dogs, ready to warn of any danger. For example a smell of burning or a noise will generally be associated with the safety aspect of a manual application rather than with its actual execution.

This is why the computer will tend to transmit signals to the operator that involve the senses of sight and touch. In addition to these senses, hearing should be added because in some cases, the computer, acting as second operator, can synthesis oral language and even understand oral commands (Pierrel, 1975; Moreau, 1981; Dupeyrat, 1981).

Although the different ways of handling visual signals are relatively easy to define, the operator looks at images on a VDU (the composition of the image is another problem altogether), or watches variations in intensity of other indicators (eg flashing or simple display of lights), touch can be integrated in more varied ways. This is because the entire surface of the operator's skin is sensitive to contact, although its characteristics vary (Catton, 1970; Hensel, 1974). This generalization of the sense of touch is used to advantage in teleoperation systems for handicapped people. Areas of the body in which sensitivity is still present are used to receive the signals (Gaillard, 1981, 1983; Gaillard and Quetin, 1982; Quetin, 1982). Signals between the skin and an external system can be divided into two types:

1. A general type, concerning the perception of force feedback transmitted to the operator through the slave arm, via the computer which may transform it: in this case, the force of reaction is distributed unevenly over the whole surface of the hand in contact with the master system. It is transmitted to the various muscles involved in the movement being carried out and it is above all the internal sensing ability of these muscles to the whole articular chain that allows the operator to interpret the force of reaction. This phenomenon is discussed in Part 3, which is concerned with the problems of the man-machine interface (see Section 3.2.2.1).
2. A localized type, concerning the epidermal and possibly the dermal sensors (see Volume 2 of this series entitled *Interaction with the Environment: Robot Sensors and Sensing*): the signals can be linked to:
   - the pressure exerted on the skin by the excitator(s);
   - the frequency of the contact(s) (Barbier *et al.*, 1981);
   - the spatial distribution of the contact points (surface area of each electrode, distance between electrodes, spatial distribution). An example of this can be seen in the machines used for reading by tactile coding of Braille (eg the Optacon system at the Stanford Research Institute, and the Delta system by the ENSEEIHT and the LSI, Toulouse) (Conter and Valobra, 1980; Lopez Krahe, 1979).

2.3.3 The choice of data to be transmitted to the operator

The problem of choosing specific data can be avoided by stating generally that 'everything useful to the accomplishment of the task' should be transmitted. But this will depend on the system, the task and the operator, which means that, once again, it is impossible to formalize the problem completely. There is no alternative, in the context of this study, but to form hypotheses and make heuristic choices. Let it be assumed, first, that the operator in question is more or less normal, but has a high level of experience of the teleoperation system. The tasks can concern the displacement of objects, assembling parts and trajectory tracking in as precise a way as possible etc. The computer is connected to all the standard types of sensors available. It must now be decided what data should be transmitted to the operator.

This data can be divided into three main categories, which can be referred to as defining the state of the task, the state of the task environment in the slave part and the state of the resources.

Instantaneous state of the task

This concerns parameters defining the instantaneous state of the relationship between the tool and its immediate finalized environment. The use of the qualification finalized means that with one gripper in two successive experiments, the same physical relationships with two identical objects may have an entirely different meaning at the decision making level if, in the first experiment, for example, the aim is to grasp a given object but in the second experiment, the same object is seen as being an obstacle.

In practice, physical sensors provide measurements concerning the physical relationships between the tool and its immediate environment. But if the analysis of the situation is not the work of a single operator, and is carried out by the computer as well, strategic advice may be sent to the operator. The physical relationships are basically geometrical values (position and orientation of the gripper with respect to the environment), forces and torques (exerted between the gripper and the environment).

State of the task environment in the slave part

The instantaneous task is one carried out in a sampling period and, therefore, concerns a very small volume encircling the terminal device of the slave arm. The rest of the volume that can be attained by the slave arm forms the instantaneous task environment. The same type of information as before, giving the relationships between this volume and and the gripper, should be available. The relationships are simply geometrical and should allow a strategy for action to be developed.

State of the resources

The term resources is taken to signify everything which qualifies as the teleoperation system, but is not the tool (end effector) or its near or distant environment in the slave environment. The resources are, therefore, made up of the master arm, the slave arm (up to the gripper), the computer and any other system present. The state of the resources can be characterized by all the data allowing action to be taken on the end effector which, itself, acts on the environment. This may include:

1. the configurations of both arms and the forces exerted (in the articulations for action to be taken, elsewhere for safety);
2. the state of effective operation of the computer and the arms;
3. data on what is allowed and forbidden;
4. distances from each articulation to its limits;
5. strategic data;
6. etc.

2.3.4 Whether, and in what form, to present visual signals to the operator

The problem is as follows: once the nature of the signals to be transmitted has been defined (eg to describe an instantaneous task), the forces exerted on the gripper and the geometrical relationship between the gripper and the immediate environment must be found. Mathematically, that is in a form that can be coded and transmitted, this means that a list must be made containing all the points of the gripper and the forces and distances to be associated with these and the corresponding points in the immediate environment. This is clearly not possible from the point of view of:

1. the measurement involved (which sensors to use);
2. the number of values to be stored;
3. the operator, to whom this list of unusable figures would be transmitted.

It is known, moreover, that in practice the operator only uses a limited number of synthetic values, whatever the problem in hand (Borghese, 1981). Two practicable solutions are possible.

2.3.4.1 REDUCING THE LIST USING AN APPROXIMATE MODEL

As far as possible this infinite list is reduced using a model with very little data and representing it approximately. Using the same example as before, the set of forces will be represented by those exerted at specific points, accessible for measurement. Figure 2.2.2.(1) shows that the forces can be summarized by those shown at A, A′, B and the torques about F_x, F_y, F_z. In the same way, the geometrical relationship between the gripper and the environment can be represented by the orientation of a trihedron linked to the gripper, with respect to the fixed set of coordinate axes familiar to the operator, and by a number of distances (and their direction) between specific points on the gripper and in the environment (points equipped with proximity sensors).

This reduced and observable list will generally contain the basic pertinent data. It may be put to two uses:

1. It can be communicated to the operator as it is in the form of a print out: despite the data compression, it requires quite a feat of imagination to picture such a scene from the list.
2. It can be communicated to the computer which:
   – generates strategic data from the operator on the basis of its analysis;
   – or generates a simple representation of the scene, for example, three-dimensionally on a VDU screen available to the operator. The operator will be able to comprehend the situation immediately, if the choice of graphic representation is suitable. It must be noted that the operator is performing a dual action by (a) understanding the situation itself and (b) generating commands to develop the situation as a function of his comprehension of the scene. The graphic representation is, therefore, not ad lib; the choices to be made in relation to this should be considered as part of the problem of the man-machine interface as a whole (Stark, 1966).

2.3.4.2 FILMING THE GRIPPER AND ITS IMMEDIATE ENVIRONMENT

For certain types of data (basically linked to the spatial structure), this list is not suitable for reduction (because there is no way of reducing it) or for storing as proposed in Section 2.3.4.1. The solution consists, therefore, of taking a picture with, for example, a video camera, showing the gripper and its immediate environment. The picture too may have two possible main uses:

1. It may be directly available to the operator on a VDU screen, leaving the operator to interpret the picture himself. This at first appears to be a simple procedure, but there are a number of problems caused by:
   – the absence of relief;
   – the faults in the picture;
   – the fact that it contains a certain amount of useless data constituting a background noise from which the operator must take the data he needs (instantaneously);
   – the need to follow the scene which is mobile because the gripper moves etc.
2. It may be sent to the computer which carries out image processing and may either calculate the characteristic values and communicate them to the operator (Bęjczy, 1977) or extract the main contours and display them in such a way as to produce a sort of skeleton from which the redundant noise is removed, or else reconstruct

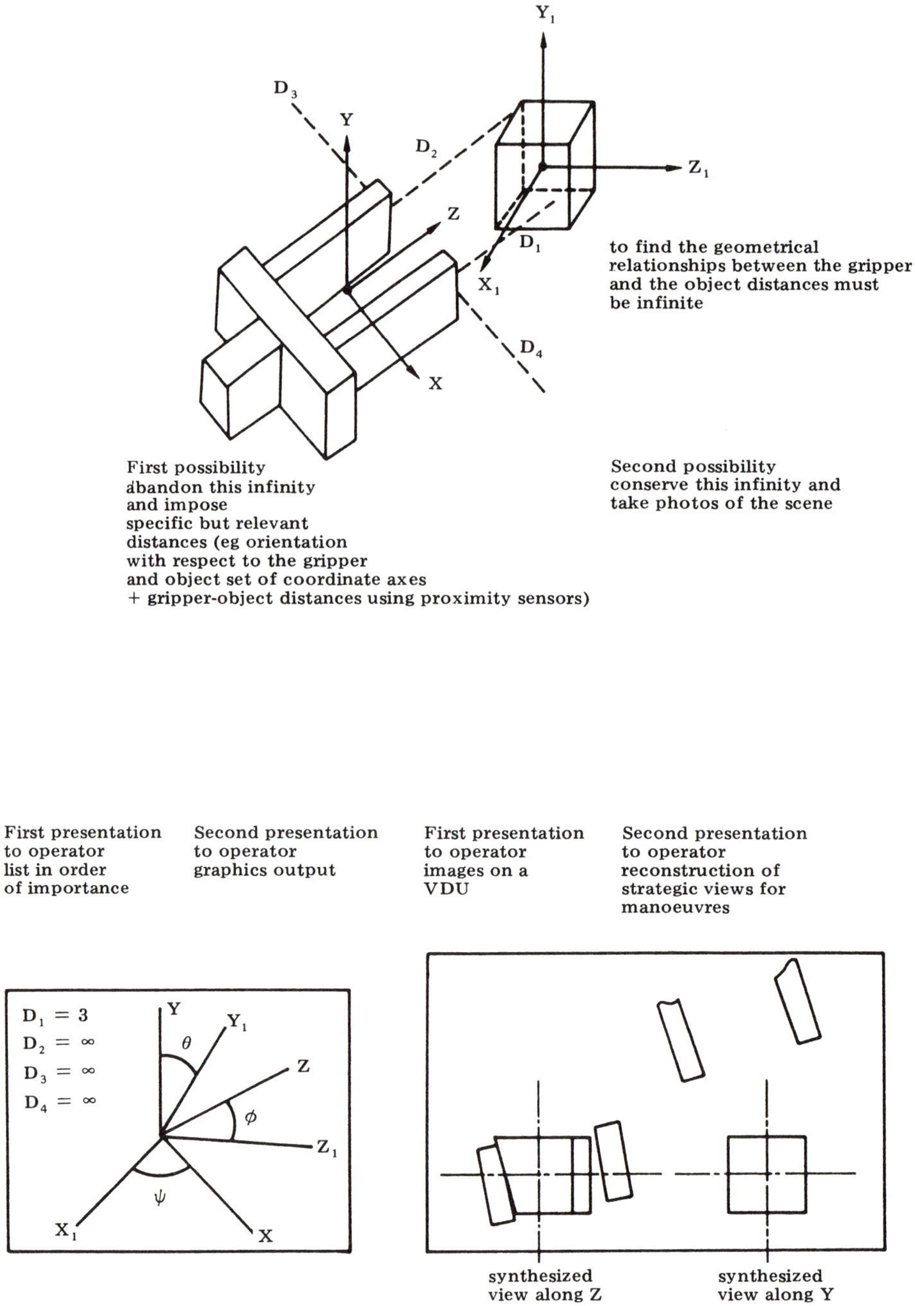

Figure 2.3.4.(1) *Filming the gripper and its immediate environment*

the image in two (or three) dimensions corresponding to the most useful view points for the operator [see Figure 2.3.4.(1)].

However, there are two basic problems:

1. On the task side, to know what data parameters are important for the operator and/or the computer, so that progress can be made in completing the task: these parameters have two characteristics:
   – it must be possible to observe, and therefore measure them;
   – they can form a variety of series depending on whether their interpretation is to be carried out by the operator or by the computer. The computer makes its recognition on the basis of the calculations requested in its predetermined programs, but the operator works in an entirely different way, the details of which are still not fully known (Gibson, 1966). (It is clear that recognition of a man from a photograph does not take place in the same way in man and in computers equipped with suitable programs.)
2. On the operator side, data presentation on a suitable medium is almost as important as the data itself for the performance of a teleoperation system. From this point of view, researchers are handicapped by the fact that little is known about the human visual recognition system (and even less about its dynamic aspect, which allows instantaneous action concerning the request for the desired modification to the image). Various solutions to this problem have been attempted, and are based on two basic methods. The first consists of avoiding the problem by attempting to re-establish the condition of normal vision by artificial means. To this end, systems for display of the scene in three dimensions, in colour, from different view points, and respecting the size of objects etc have been developed (Amos and Wang, 1978; Charles, 1979; Foote and Dompe, 1978). Experiments have shown that these methods are still not perfect, but notable among the attempts at reconstituting relief in the image, is the Honeywell system. The operator wears glasses, each lens of which becomes opaque in sequence with display of a right and left image on a VDU screen, provided by a stereoscopic camera (60 images per second) (Berry and Rice, 1978) (see Section 1.5.6).

   The second method consists of trying to replace the relevant values in the images in as simple a way as possible (stylized), and in relation to the sets of coordinate axes inside which the operator can situate himself at control level. Although a simple perspective view is, therefore, frequently used to help the operator comprehend the situation, views made up of cross-sections parallel to the planes, and listed in terms of the sets of coordinate axes along the axes of which movement can be made, are also helpful in exerting control [see Figure 2.3.4.(1)]. The same technique can be used to represent not only the geometrical relationships, but also the forces involved (Gaillard, 1983; Bejczy and Dotson, 1982) or amplitudes of any value. The quality of such systems cannot be calculated in advance, and measurements can only be made on existing systems, allowing qualification of the various choices of representation tested, on a relative basis.

2.3.5 Redundancy and complementarity of vision and touch

The respective roles of touch (or more precisely, the sense of force) and vision have already been discussed in Volume 2 of the series entitled *Interaction with the Environment: Robot Sensors and Sensing.* Touch itself as a means of identifying geometrical forms is made redundant by vision. This complementarity can be summed up:

1. vision allows the distance between the current state and the objective to be determined;
2. the sense of force allows the identification trajectory to be followed at each instant, to make this distance decrease, and allows the operator to control the progress of the trajectory.

A typical example of this type of complementarity can be seen in assembly work involving the insertion of a cylinder into a bore hole with low tolerance, using a remote manipulator. On the screen the images provided by the video camera(s) show the gripper, the cylinder and the bore hole. An idea of the distance between the current state and the objective can be easily gained by simply looking at the screen. If the extreme point of the cylinder is at the entrance of the hole but slightly inclined, the use of vision allows the operator to see that the angle of the cylinder must be corrected.

Experimentation has shown, however, that without force feedback, assembly takes a very long time and correction is difficult in case of mistakes. With force feedback, the operator can detect the forces exerted between the cylinder and bore hole, and can easily complete the task by making small movements and by predicting, at each instant, the correct following microdisplacement. This, of course, involves a certain amount of training because the mechanical structure as a whole constitutes, because of the threshold of friction, flexibility and, above all, backlash, a filter between the force to be detected and the force felt. (The training is precisely what allows the operator to detect the nature of the filter.) Figure 2.3.5.(1) illustrates Bejczy's approach at the Jet Propulsion Laboratory (JPL) to the control of bilateral servo manipulators (generalized master-slave type) and the visual display of contact force data.

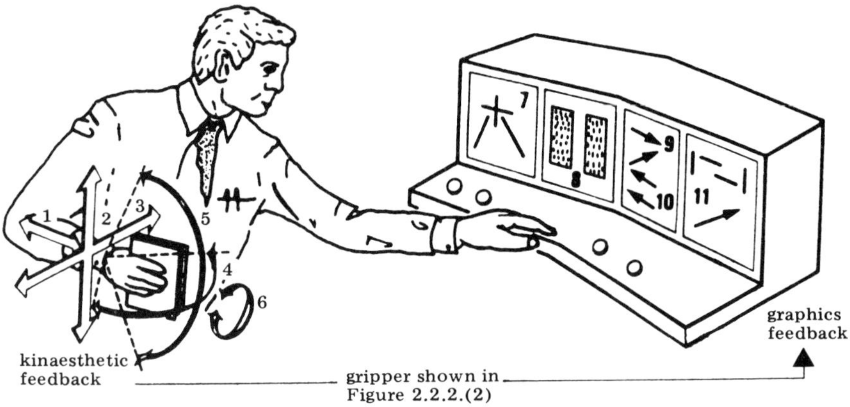

Figure 2.3.5.(1) *Bejczy's approach to the control of bilateral servo manipulators and the visual display of kinaesthetic and force feedback data*

Controls
translation 1: in-out
translation 2: up-down
translation 3: right-left
rotation 4: azimuth
rotation 5: site
rotation 6: roll

Visual display
7: proximity
direction
distances
8: contact (between the object and the gripper)
and pressure distribution
9: force
10: torque
11: stress
12: format knob

[from the Jet Propulsion Laboratory, see also Figure 1.7.11.(1)]

Conclusions

This chapter has been kept short deliberately because the intention is simply to evoke the criteria and problems of data transmission to the operator, and to draw attention to the fact that the data function as provided by the computer presupposes thorough knowledge of the human element, particularly in connection with the way in which images are comprehended and, more generally, the human faculty for integration and response to sensory stimuli. There remains a great deal of progress to be made in order to master this field, and so far human operator models are greatly simplified [(elementary tasks, with one or two inputs and outputs) (Malvache, 1981)]. The complexity of a teleoperation system has been shown here, as has that of the operator-teleoperator relationship. This complexity only increases with reference to the individual variations in human behaviour and the extremely variable functional handicaps found in medical remote manipulation (Guittet, 1985II). Consequently, the major choices will be made heuristically for a long time to come and optimization of the nature of data and its presentation to the operator will remain highly localized. This area of study is still wide open.

Part 3
Performance and the man-machine interface

Chapter 3.1

Performance evaluation of teleoperation systems

3.1.1 How can the performance of a teleoperation system be defined and assessed?

These questions were answered for industrial robots in Volume 1 of this series entitled *Modelling and Control.* To the best of our knowledge, an international group chaired by Thomas Sheridan in 1975 is the only group to have considered these problems of performance evaluation; their report (Sheridan, 1975) is cited several times throughout this volume. One of the authors of this volume also took part in the study.

Overall performance can be described as the quality (precision) with which applications, run on a system manned by a human operator, are carried out. This is a very general definition. It is impossible to be more specific, since the condition of quality by definition opposes the condition of quantity, which alone interests scientists. Moreover, it presents many questions because:

1. The applications planned for a teleoperation system are ill-defined. They may, in principle (as for robots), be varied and they may not be included in the system's specifications.
2. A teleoperation system includes a human element. Each operator has his own level of skill and speed of learning. Is there such a thing as a 'standard' person?
3. The measurement of performance only makes sense in relation to a reference. The first idea consists of using the characteristics of a task carried out bare handed as the reference. However, if the teleoperation system is, for example, to position a load weighing 200 kg, this reference becomes unsuitable. In such an example should the reference be a scaled-down version of the task? How can the size and force scales be related?
4. The analysis of the complexity of teleoperation systems implies that, even assuming that there exists a standard human operator, performance is a function of many parameters: hierarchy, dependence and even the identification of these parameters. (Consider that it only takes a small deformity or lack of lubrication in a simple mechanical device to reduce performance standards

considerably. If this remains unobserved it can result in a serious error in the appreciation of real performance of the system.)

Thus, anything can contribute to the greatest confusion or the greatest risk in measuring the performance of a teleoperation system. In addition, comparisons made between existing systems are often subjective or based on very specific points, the relevance of which to the overall comparison being not known.

3.1.2 Qualitative performance: manipulator properties

In 1967 Johnsen drew up an index of 32 properties on which the performance of a mechanical or motorized teleoperation system could be judged. A modified form of the index is used here, and the properties are classified into five categories (Johnsen and Corliss, 1967).

3.1.2.1 FACTORS THAT RELATE TO THE MECHANICS AND/OR MOTORIZATION

Coverage or reachable space of the slave system

This space is characterized by two elements (see Chapter 1.3 and Volume 1 of this series entitled *Modelling and Control*):

1. the *volume* in which the tip of the end effector can be moved;
2. for each of the points in this volume, the *orientation* (or shape) of the end effector (that is the possible amplitude of three rotations of the gripper about the three intersecting axes that relate to it).

The reachable space is also characterized by the set of configurations that the arm can adopt. This has given rise to a classification of the zones within manipulator coverage (Vertut, 1973I, II, III, IV). For an effective volume of one or other of the zones mentioned above, a parasite volume resulting from the manipulator itself and particularly its motor, or the extremity of its large transporter, limits the use of the manipulator in the task area. The *volumetric efficiency quotient* must therefore be taken into account (Fraize and Vertut, 1984).

The effective load and available force

This is the weight that can be displaced throughout the reachable space. It should not be confused with the *payload*, which is the maximum weight that the manipulator can handle in particularly favourable configurations.

The ratio between the effective load and the maximum force characterizes the *isotropy* of the manipulator, mentioned in Chapter 1.5.

Moreover, the concept of effective load, in the case of motorized manipulators, may be more clearly explained in the sense that it is the load that can be permanently applied to the end of the robot arm under normal conditions of the actuators, or the maximum load held for a limited time. Finally the concept of effective load as the mass and weight in the earth's gravitational field changes its meaning in outer space teleoperation and even in some underwater operations, taking account of the effect of the Archimedean force.

The grasping force: If a manipulator is equipped with a gripper, it must follow that the grasping force must be compatible with maintaining the effective load (weight) in the gripper and/or with the force exerted on the object (static or dynamic action).

In remote manipulation, there is more than the simple transfer of objects to be considered. Using the prehensor, end effectors can be held (eg tools or machines such as a hammer, drill, chisel, wrench) so as to run applications in the environment. The environment exerts forces and torques on the tool, which must be held firmly in the gripper, with a combined force for simple opposition to gravity. This is why the grasping force is a parameter that participates directly in manipulator performance.

The ratio of the effective load to the mass of the arm (or mass ratio)

This ratio is a very important property of manipulators mounted on large transporters, and is also essential when articulated systems lift all or part of their own weight, as is the case with articulated legged systems. Although this ratio characterizes the quality of the mechanics, it is not an indication of its cost, because very costly technical additions are required to improve this factor to any extent. In normal configurations where the slave manipulator is mounted on a large transporter and/or vehicle, reduction in mass justifies the additional expense because of the economy which is made in the other components as a whole. The improvement is, in particular, the key to the development of teleoperation in the maintenance of nuclear reactors.

The number of degrees of freedom

Without formalizing the relationship between the number of DOF of a manipulator and the classes of tasks that it can execute, it can be seen that the more DOF there are, the greater the versatility of the system. The minimal number providing geometrical versatility, and allowing any direction to be taken within a certain space, is six.

Coupling between joints

Transmission between motors and joints presents a serious mechanical problem. It is also commonly found that either a motor participates in driving several joints or that there is mechanical coupling between two joints. In general, the coupling principle is a drawback (ie in trying to act on one DOF, another is activated) but may, in certain applications, become an advantage. This is known as *coordination of the DOF*. Parallelogram coupling, with the articulations making use of differentials, has advantages and is discussed in Section 1.5.3.

Compactness

This is the ability of the slave to adopt a 'stowed' configuration, allowing the manipulator to occupy the smallest possible volume for transport. This is very useful for manipulators on submarines or satellites, but these manipulators are also used in the nuclear industry, where it is often necessary for the manipulator to be able to move in and out of its housing, thus minimizing contamination (Fou, 1982).

Torque available on the wrist

One of the common tasks in teleoperation is that of tightening or untightening bolts, nuts or screws. It is, therefore, important that the rotation torque of the gripper about itself should be sufficiently large to allow these operations.

Inertia

This is a measure of the difficulty in accelerating or braking. It may cause displacements and oscillations when tracking a target. The origin may be mechanical but it can also be electronic (delays). Above a certain level it can become tiring for the operator.

Friction

Depending on the type of friction (dry or viscous), its effect on the operator will be different. *Dry friction* will be manifested as thresholds to be overcome and, depending on their level, will have an influence on movement precision (if the operator is surprised by the sudden movement following each threshold), and on operator fatigue. *Viscous friction*, by virtue of its linearity, is not in opposition to the effective control of movement but the operator uses up energy proportionally to the velocity, as pure loss.

Backlash

This exists in most gearing systems, and it is manifested in hysteresis. If the gearing is made to move first in one direction and then the other, for example with a command using potentiometer, there is a dead zone in which the gears do not move. In computer control this is very intrusive, because the phenomenon is not always stationary, and it will have an effect on precision.

Flexibility

Experiments have shown that flexibility, considered as a fault when a command is given for precise positioning, can be useful (on condition that the flexibility remains within certain limits) in a mechanical master-slave system. The flexibility of the arm can help manipulation, particularly for insertion or assembly operations. The operator controls the system using his force and the manipulator has the advantage of being able to flex slightly to allow insertion despite slight errors in positioning. A very rigid manipulator would soon tire the operator under the same circumstances, and greatly extend the execution time of delicate tasks.

Compliance between possible movements and the movements required for the task

This property relates to the reachable space and the DOF, that is, the geometrical structure of the manipulator and the range of each joint. It can be possible for the manipulator to reach a desired point within the work volume, but physically impossible to orient correctly the gripper at this point in relation to the required task. This property is known as *compliance* (defined by *The Concise Oxford Dictionary* as Action in accordance with request, command etc). It should not be confused, although it has the same origin, with *mechanical compliance* discussed in Volume 2 of this series entitled *Interaction with the Environment: Robot Sensors and Sensing* in connection with robots, which evokes the idea of flexibility of a segment, but calculated so as to attain an objective, for example, insertion.

3.1.2.2 FACTORS THAT RELATE TO CONTROL

Velocity and acceleration

In mechanical master-slave manipulators or bilateral servo manipulators, the velocity of an operation is dictated by the speed with which the operator moves the master arm and by the velocity limitations inherent in the slave arm. If it is assumed that the manipulator or its servo-system does not limit the velocity required by the operator, then friction and inertia may oppose it.

Velocity is an important property when the operator acts upon a master part that is very different in structure to the slave arm (eg push-button system) or when the computer ensures that the displacements are carried out automatically. Systems can be designed to move at high velocities, but precision is always reduced. There is, in fact, less interest in the instantaneous velocity than in the execution time for the task (see Volume 1 of this series entitled *Modelling and Control*), and this involves periods of high and low velocity. In consequence, maximum velocity, whether considered for each DOF or in the task space, is a property which has little meaning outside its context, which is the control of the velocity, and depends to a large extent on the operator.

Positional precision

Although this depends on mechanical factors, such as flexibility or backlash, it is also dependent on all the control components (eg sensors, servocontrol loops). The precision considered here concerns that of the actual configuration of the slave arm with respect to a configuration requested by the control system. This is manifested in the task space by positional precision of the end effector and its orientation precision. The importance of this property depends largely on the mode of control chosen:

1. When teleoperation is mechanically or electrically controlled the absolute precision in the task set of coordinate axes is not important. The operator controls the system with the help of the relative position of the end effector, judged by eye. The operator analyses the situation in terms of 'closer', 'lower', 'forward' etc.
2. In periods when the computer is substituted partly or totally for the operator in a manoeuvre, two cases must be considered:
   - The slave arm is equipped with suitable sensors and the commands are given in relation to the target, that is, the environment. In this situation, absolute precision with respect to the fixed set of coordinate axes is not important. Distance is measured, for example, and adaptive control is used to act on the arm until the distance reaches the desired value within the allowed tolerance (Coiffet and Rives, 1980).
   - The slave arm is a first generation robot, that is to say, it operates in an open loop with respect to the environment. Under these conditions, there is no alternative other than to send the arm into a configuration calculated with respect to the fixed set of coordinate axes. This is the only case in which positional precision becomes a significant merit factor.

This precision is not only obtained by rigidity and the resolution in the system. When its deflection can be modelled, the *static deflection model* allows precision to be obtained in rigid systems at very low velocity.

Positional resolution

This concerns the smallest movement the system can perform, which can sometimes ruin the concept of precision. In teleoperation, resolution is more important than precision, and is derived from the friction and servocontrol gain. The smallest displacement requires the friction factor to be overcome, hence its value is in inverse proportions to the gain. This demonstrates the great importance of having low levels of friction in teleoperation systems.

Stability

This is the capacity to lead the arm from one configuration to another smoothly (without jerking) and to maintain the desired configuration without any oscillations being produced. This depends largely on the adjustment of the servo systems, particularly in bilateral systems where a loop is closed by the operator's hand.

Drift

Drift is particularly noticeable in push-button control systems and describes a continuous movement at very low velocity of the slave arm away from the desired configuration. When there is velocity servocontrol it corresponds to the fact that zero velocity is not easily obtained, and this may result from slight reversibility in the system, or hydraulic leakage when the system is left without supply at rest. This is why many manipulators have brakes to block the actuators when the power supply is cut-off.

3.1.2.3 FACTORS THAT RELATE TO THE HUMAN OPERATOR

Sensitivity

This is often paired with stability, and concerns the ratio between a micromovement in the master system and the corresponding movement in the slave arm. An oversensitive teleoperator will be very difficult to control manually in terms of velocity, but perfect for positional control. This illustrates the need for compromise between backlash, friction, servocontrol gain, and the spatial displacement ratio, to achieve improved ergonomics. Unfortunately, since work on these subjects is only just starting (eg the ARA project), these points are not fully understood, and are linked to the quality of controllability and feedback, particularly force feedback, of the system.

Quality of feedback signals to the operator

This is an important property, which is difficult to define, but of the greatest interest since it allows the operator to develop a control strategy. The importance of visual feedback is well known, but force feedback (which cannot be envisaged with push-button control or computer peripherals) is also very important.

Control station design

To run a task under the best possible conditions, the operator must not tire mentally or physically too quickly. To avoid these effects the control system must be adaptable to the human form, both with respect to posture and the movements and forces to be exerted. In addition, the feedback signals received by the operator must not exceed a certain rate and should be relevant to the application. At the same time they must be in a form that minimizes the amount of thought and attention required from the operator.

This is all the more important in cases where interference may originate from movements at the control station. For example, it may become necessary to apply a stabilizing effect to minimize these influences on board ship.

Functions of the computer

These functions have been mentioned in Part 2, and can range from simple substitution of the operator, to work in parallel, to informing the operator, thus involving the man-machine interface.

Operator qualification

This concerns the relationship between the time taken to train an operator how to use a teleoperation system and his intelligence quotient (IQ), dexterity and knowledge of the computers and machines concerned. The nearer the level of operator qualification is to being 'standard', the better the teleoperation system, all other things being equal.

3.1.2.4 FACTORS THAT RELATE TO SAFETY

Robustness

This is the capacity of a part of or the complete electromechanical or hydromechanical system to continue to function under abnormal conditions. A collision at high speed between a manipulator and its environment is always a possibility, as is a short-circuit etc.

In this way an electrically controlled manipulator may, due to an amplifier fault, for example, function in an abnormal way with maximum torque and velocity, and as a result collide with its limits or the environment. Robustness may be a result of weight reduction, which minimizes the energy involved in such collisions because of reduced inertia. On the other hand, the request for excessive force is manifested by an increasing error signal, but the system itself can resist these overloads. The system will be more robust if it can resist a number of incorrect types of treatment.

Resistance to environmental factors

This is one aspect of robustness, and it relates to the physical properties of the environment in which the (slave) manipulator is placed (eg a vacuum, sea water, outer space, furnaces, X and γ radiation sources). The simpler the teleoperation system, the better it will perform from this point of view.

The mechanics will resist corrosion well, except when the problems relate to the lubrication, and the step-down gears should be protected from the environment. In the same way the electronic components and sensors must also be protected. It is clear that the weak points in the resistance to the environment are the interfaces between the mechanical and electronic parts, in particular the connectors.

Reliability

This is defined by the probability of there being a breakdown during a given interval of time. It is related to the complexity of the system; mechanical manipulators may be more reliable than computer-aided manipulators, because of the additional failures that may result from more complex systems. In practice, servocontrol imposes the automatic limitation of force that contributes so much to the avoidance of manipulator overloading that the resulting mechanical reliability is greatly improved as compared with a mechanical master-slave manipulator (controlled manually by the operator), in which reliability is reduced because of overloading abuse by the operator. The establishment of a hierarchy of failures is an important element in the problem of reliability, since the importance of the task to be carried out may be such that safe running is seen as being a priority. Depending on the type of failure it should be possible to continue to carry out a task, although possibly less easily. The mode of operation used in such circumstances is said to be degraded (Dombre and Liégeois, 1979; Bertrand *et al.*, 1982; Liégeois, 1982), which is less serious than complete immobilization of the system.

Immobilization of an intervention system in its workplace is unacceptable in many cases, and the system must also contain sufficient redundancy at least to allow its removal without execution of the task.

Maintainability or repairability

This is the facility of access to the different components, and, in particular, to transmissions and motors in order to allow repairs to be carried out. It should be considered both from the point of view of mechanical and electronic maintenance. A failure in the electromechanics of a manipulator on board an unmanned satellite is virtually irrepairable but this is not the case with electronics in which some circuits can be duplicated, with automatic switching from one to the other or on command from the master station. This example shows how the electronics can be equipped with self-repairability, but for mechanical or electromechanical parts an intermediate stage must be envisaged, involving remote repair using another manipulator. This is impossible in outer space, but it could apply under water or on land, and particularly in a nuclear situation. There are standard techniques for interchanging modular electromechanical components.

Self-repairability

This relates to an aspect of repairability which may or may not involve human intervention.

1. Suppose a failure occurs in the slave arm and a mechanical part has to be changed, involving a certain amount of dismantling. Current practice in remote manipulation involves the operator working with both hands, using two master arms which control two slave arms working in cooperation. If one arm is faulty, repair can be carried out with the other, and the master-slave system is said to be *self-repairable.* The term, as it is used here, indicates redundancy of the arms and human intervention in the system.
2. This type of self-repairability does not require continuous human intervention, can even occur without the operator's knowledge (software failures) or it may only require an elementary command from the operator when the diagnosis is entirely automatic. For example, the computer indicates to the operator that electronic card number 18 is functioning abnormally. The operator has simply to instruct that card 18a is to be used, or if the system is designed for it, that card 18 should be replaced. This is a scenario planned for the maintenance of satellites using manipulators, but to date these systems only exist in theory.

Safeguard capacity

Just as any computer has a facility to store a certain amount of data for the continuation of work following a failure in a particular section, a teleoperator which is subject to a power failure or fault in the control circuit must be able to hold its arm configuration when the failure occurs, and in particular it must not drop the load. This ensures that the load is not lost or damaged.

Self-protection

To protect an electric motor, a fuse or circuit breaker must be installed. Although these cannot always be reset remotely, a current overload will lead to protection against severe damage. Switching on the circuit must be controllable or automatic in order to qualify as being self-protective. This property contributes greatly to the life of the remote manipulator. In standard mechanical models, this self-protection is provided by end switches. In computerized teleoperation, self-protection is connected to the prevention function entrusted to the computer. Self-protection includes two aspects mentioned previously: an informative aspect in which the operator is warned of danger, and an active aspect in which the operator's action is automatically opposed physically.

Availability

In cases where the system is not in constant or frequent use (eg for intervention devices), it is very important that the system will start up as required. This can range from simple immediate operation to operation following a countdown. Here also the computer is a great help.

3.1.2.5 OTHER FACTORS

Cost

Unlike the industrial robot, the teleoperator is not intended to operate on a production line. As a result only small numbers of teleoperators have been manufactured, and they are used as safety equipment, usually kept in reserve for an emergency. All this leads to high prices, and the impression that they are not much used, which appears not to justify the price. It is not unusual to work with two arms. In spite of developing master devices with reduced volume (which is more economic than the slave arm built to withstand the hostile environment) such a two arm system requires the mechanical production of four manipulators. To this can be added, for large systems, a large transporter, possibly a vehicle and a non-cable controlled communication system etc. The introduction of permanent computer aid involves the addition of sensors and peripherals, which certainly increase the cost. On the other hand, all the previous points mentioned in relation to increased reliability and operational safety are of benefit (including the relief feeling for the operator, the possible reduction in the number of technicians required for a system and, finally, the most important point of all: the greater speed of execution for a task that does not have to be fully defined in advance). This indicates the direction in which maintenance teleoperation will take: with reasonably frequent use, and rapid performance, teleoperation will improve accessibility to facilities and become irreplaceable in cases of serious accidents. This modifies the cost notion and furthermore may radically alter the design of new facilities. For example, such is the philosophy of the Remotec nuclear fuel reprocessing concept (White, 1980).

Integrated power and energy efficiency

Electric or hydraulic actuators often require power sources of several kilowatts in the slave environment. This source adds to the size and weight of the slave system, and it could be a severe drawback in a submarine, in launching a satellite or in an autonomous land-based vehicle. This point is enhanced by the fact that teleoperation is designed to continue over a period that allows the task to be completed (with successive operators at the control station, if necessary). The effort to improve the energy efficiency of the global system has a major effect on the final weight of the slave system. For the near future, this may even affect the feasibility of some systems.

Aesthetics

This aspect only arises in medical applications, particularly for prosthetic devices. It is important that handicapped people should not be distinguishable from the able bodied, and a prosthesis should not only look like a real limb (which once again poses the problems of volume and mass, for example, in the components of an artificial hand), but it should move in the same way as a real limb (silently and without jerking).

The aesthetic problem also arises for telethesis (or computerized manipulator remotely controlled by the handicapped person), but this time the problem is closely related to standard industrial and household aesthetic criteria.

3.1.3 Towards quantitative performance measurement

The important properties considered above are numerous; some are contradictory, others do not apply to all teleoperation systems; some are difficult to quantify. They are not all of equal importance and the hierarchy often depends on the application; they are not very independent, and throw into question parameters that are sometimes not observable and sometimes even mathematically indefinable. How, therefore, can the overall performance of a teleoperation system be characterized by numbers?

Given the complexity of the problem, one might expect there to be many approaches, and this is indeed the case.

3.1.3.1 MEASUREMENT BY CLASSES OF MERIT FACTOR

This can be compared with the method used for car performance, in which various aspects relating to use are considered: road holding, speed, fuel consumption, comfort etc; all elements which are never considered altogether because the user will attach a coefficient of his choice to each one, and this hierarchy varies considerably from one driver to another: one point may be vital, the others may be important or not even taken into consideration.

For better understanding of the relativity of performance evaluation, let us consider a system evaluated in terms of five properties or merit factors: A, B, C, D, E, which can be precisely quantified and marked out of 10. The user (or buyer) is assumed to understand the notation fully, and his wishes are clearly expressed: he wants $A > 8$, $B > 5$, $C > 5$ and D and E are not of interest to him.

A system with the following performance: $A=B=C=D=E= 7$ would not be a good system by his evaluation, although the system is intrinsically good or good overall. This justifies the evaluation of the teleoperation system by diversified properties or merit factors.

Ozaki *et al.* (1977) proposed a performance index of 25 classes, which is reminiscent of that devised by Johnsen (although this is not cited), and for each one he lists the parameters that should be measured (but not the method). There are four categories of merit factor:

1. General merit factors:
   (a) Into what class (kinematic) does the robot or manipulator fall?
   (b) What are its dimensions?
   (c) For what type of intervention is it intended (eg for maintenance, dangerous situations, production lines)?
   (d) What are the average and maximum velocities of each articuation?
   (e) What are the power requirements?
   (f) How can the system be assembled on site?
   (g) What environmental constraints must be respected?
2. Merit factors that relate to the mechanism:
   (a) What are the modes of control (eg sequential control, training mode)?

   (b) What is the mechanical construction like (eg axis 1 is actuated by a hydraulic linear actuator with a 300 mm stroke)?
   (c) How are the mechanical stops defined (eg fixed, programmable)?
   (d) Characterization of programming mode (if applicable)? How is the robot positioned? How is its velocity controlled? What is its memory capacity?
   (e) What internal sensors does it have?
   (f) What safety measures are there?
   (g) Can the system be interfaced with other external systems, and how?
   (h) What are the possibilities of the end effector?
3. Functional merit factors:
   (a) Positional precision by manual control (eg training period for a robot).
   (b) Positional precision in automatic mode.
   (c) Transfer time between two configurations.
   (d) Duration of repetitive cycle.
   (e) Velocity to precision.
   (f) Possibility for acceleration and braking for safety.
   (g) Average and maximum effective load.
   (h) Positional drift.
   (i) Braking distance until stationary for emergency stop.
4. Miscellaneous
   (a) Reliability (average time between breakdowns).
   (b) Maintenance.

A more rational approach has been adopted by Warnecke and Brodbeck (1977), who devised a test bed for robots or manipulators. They are also interested in the following four classes.

1. Merit factors concerning the geometrical dimensions:
   (a) Work space; range of each articulation (translation or rotation).
   (b) Positional precision in static and dynamic mode, and in the long term.
   (c) Overshooting.
2. Merit factors concerning kinematics and dynamics:
   (a) Operational duration, velocity.
   (b) Maximum acceleration.
   (c) Grasping force of gripper.
   (d) Available force in the direction of movement.
3. Merit factors concerning the influence of temperature (eg of the components, fluids, environment).
4. Merit factors concerning the type of power used (eg electric, hydraulic, pneumatic) and other properties (eg noise).

3.1.3.2 MEASUREMENT OF IMPORTANT MERIT FACTORS

These properties are measured for commercial robots. Manufacturers have so far shown little interest in all merit factors and have concentrated only on precision of the final configuration of the end effector both for position and orientation, in automatic as well as in training modes. For industrial robots, precise repetition of movements is essential, but little effort has been made to measure dynamic performance and the influence of the effective load (BNM, 1981).

Manufacturers of remote manipulators, on the other hand, place emphasis on the effective load and friction (bilateral manipulators) but ignore the safety factors.

3.1.3.3 TECHNIQUES FOR MEASURING RELEVANT VALUES (TAKEN FROM INDUSTRIAL ROBOTICS)

The cube seems to have exerted a magical influence over all those engaged in the experimental testing of precision performance. The cube is fixed by one of its faces to the tip of the robot. It constitutes a male part which is to be inserted into a female cube, three faces of which are removed, transforming it into a trihedron [see Figure 3.1.3.(1)]. Three

Figure 3.1.3.(1) *Measuring cube*

faces of the trihedron are equipped with sensors (eg proximity, optical, inductive) allowing the position and orientation of the cube to be measured with respect to the trihedron, the object being to join them together perfectly.

Warnecke equipped the trihedron with three inductive sensors, which allowed small distances to be measured accurately. The parameters made to vary were:

1. direction of movement (the cube is successively moved according to the three possible axes of the trihedron);
2. load;
3. velocity;
4. position of the trihedron.

He measured static positioning by moving the robot to its final configuration and noting the error with respect to the three dimensions. Dynamic positioning was evaluated in the same way, but with the robot performing a cycle passing through the desired position ten times. The inversion of the direction of movement of the cube led to another type of error (hysteresis). Long term repeatability was evaluated using a test cycle over several hours. Finally the overshooting of the desired position and the damping of oscillation resulting from it were recorded.

The same method is used in France by the Centre for Study and Research into Machine Tools (CERMO). Renault uses a cube for static measurement, on five faces of which are optical targets, each one with a sphere 0.5 mm in diameter. These are lit up using optical fibres, and measurement is made using two theodolites.

Peugeot uses the robot to hold the female trihedron, with two proximity sensors on each face. Three test cubes are placed on a mobile platform. For the control of trajectories, three potentiometers integrated with small coils of fine wire are used with a small motor for coiling up the wire. These three wires are placed in three opposing directions and attached to the end effector of the robot. Recordings of the potentiometer signals can be used to follow the trajectory of a point in the space (BNM, 1981). Hill at SRI (SRI) and Vertut and Acquaviva-Lescoat (1983) have used this method with remote manipulators.

It is also possible to use the processing of two television images or any three-dimensional vision system, on condition that the resolution obtained is compatible with the envisaged measurement.

These forms of measurement originate from knowledge amassed in industrial robotics and are within the framework of traditional teleoperation where we have to add the following considerations.

Friction

In particular for mechanical master-slave manipulators, simple dynamometric methods allow measurement of friction in starting up unloaded manipulators. To the best of our knowledge, measurements have not been taken with loads except for the block and tackle transmission of the MA23. The counterbalancing error is also measured.

Loaded deflection

The measurement currently carried out at the CEA consists of immobilizing the lever of the master system and measuring the displacements in different directions for a given load, represented either by traction, by a weight on a cable, or the force exerted by a dynamometer with a spring. So far measurement has been carried out manually and does not constitute continuous recording as a function of the force.

Inertia

Data obtained from calculation are used, following various tests on the behaviour of mechanical master-slave manipulators with pendular oscillations. The same methods of measurement were, of course, applied to the bilateral servo manipulators and new types of measurement were developed based on the specific nature of these manipulators on which positional sensors are used around the motors (which does not supply the absolute position at the extremity), and relatively precise data on the torques in the current motors or with the use of local sensors.

Measurements that relate to the force and power consumed

Work is currently in progress at La Calhène, the manufacturer of MA23 manipulators, which is opening up interesting perspectives. It concerns tests on a task for an operator or repetitive programming, with simultaneous recording on an XY recorder for the voltage and current of each motor in the remote manipulator. In this way, continuous displacement of the recorded point is obtained during execution of the task, and this allows the instantaneous force and power to be found. By establishing the correlation of each part of the task with the corresponding recordings, it is possible to recognize the instantaneous power required for execution of a task and to compare it with the power the manipulator is capable of. Combinations of other types can be used to provide other data, for example, by using the XY recorder for position and current, the surfaces of the cycles can be used to deduce the mechanical work provided by the manipulator. This can be applied to a slave or master arm to show the work applied to the system by the operator.

In another approach developed by Peruchon and Micallef (1982), a torque and force sensor with six axes is used to measure the force applied over a period of time by the teleoperation system during execution of a given task (eg insertion). In this way, it is possible to determine, for a robot or teleoperator, the parasitic forces due to the fault in the system with respect to the forces strictly needed by the measurement of the same operation carried out manually (if the task concerned is possible: manipulated mass in the scale of masses that can be manipulated bare handed). No manual work reference with a spatial and force scale factor has so far been devised.

These dynamic operation measurements can provide a very important contribution to the complete knowledge of the state of fatigue of use of the system.

The work in progress at the CEA (Sreiff, 1983) uses very simple modes of recording giving indications on the power consumed at different force thresholds so as to give a histogram of tasks actually performed on the one hand, but above all to allow determination of the fatigue and level of use of the systems for the functions mentioned above. All these measurements are currently made on the elementary movement, by selecting the movements producing most work, which implies for work on land that these are the movements working against gravity. The use of data processing will allow access to the real vector of force. We believe that the work in progress mentioned here will find its application in the assessment, for industrial robots and particularly assembly robots, of the new and relevant properties arising from the development of teleoperation.

3.1.3.4 COMPARATIVE MEASUREMENTS TAKEN IN THE PRESENCE OR ABSENCE OF CERTAIN ELEMENTS

The techniques presented so far have not taken into consideration the influence of the human operator, which is important in teleoperation and in robotics during the teaching or programming periods.

The methodological aspect of those experiments implying involvement of human operators is developed in Sections 3.2.6 and 3.2.7, along with an overall interpretation of the results, but it is still necessary to analyse the points at which the operator intervenes, particularly with regard to the difficulties involved in the task itself, the feedback signals, or to the methods of control open to the operator.

This is why another point of view, that of the measurement of *relative performance*, is needed, but the word relative implies the use of a reference which would be both absolute and unquestionable. It appears that such a reference still does not exist and each individual must use his own in the absence of any uncontested theory and because it is easier to do so; but this often rules out any possibility of comparison between systems.

A preliminary class of references could satisfy the following reasoning. Using a given teleoperation system, a certain number of tasks assumed to be representative are to be carried out (after an adequate training period), and the changes in a certain number of parameters believed to be significant are to be measured. After this, one function or element of the teleoperation system is modified and the same tasks are repeated, with the same parameters recorded. The comparison between the two series of measurement provides comparison between the performance of the system in the first state and the second. A certain number of hypotheses have been made:

1. tasks are pertinent;
2. operator has the 'correct' dexterity (after training);
3. measured parameters are highly significant.

Kugath and Wilt (1973) at General Electric used this method, taking the flexibility (compliance) of the arm as their main variable. The tasks included tracking an object, then grasping it and transporting it to a warehousing point. The performance factor can be resumed in the measurement of an execution time which will vary with compliance. The experiments were carried out on two manipulators (E2 and CAM1400). The results of these tests were also marred by uncertainty, because for one of the two systems the operators never completed their training period. The influence of compliance, therefore, cannot be transferred to other systems without reservation, but will bring about a slight increase in duration.

The authors used the same procedure for the Handyman and the E2 manipulator to evaluate the importance of force feedback in assembly tests, then modulated the visibility conditions (four qualitative tests: good, average, poor and zero visibility). These provided performance of the type shown in Table 3.1.1. The task was to insert a peg into a hole.

visibility conditions	average time with force feedback(s)	average time without force feedback	number of attempts
good	88	122	10
average	130	236	10
poor	251	573	10
zero	1,184	impossible	

Table 3.1.1 *The importance of force feedback on performance when inserting a peg into a hole under different visibility conditions*

Table 3.1.1 confirms the importance of force feedback and good visibility in a teleoperation system. The data provided also indicate that the shortest possible time in which the operation can be performed is 88 s which seems very long compared with the time taken to carry out the same task by hand (this raises an idea for a new type of reference, which will be discussed elsewhere). Book and Hannema (1980) have made a study of the problem, based on other parameters, and have made a special arm with two DOF for tracking tasks, with scale 1 positional control. The study concerned the variation in execution time for this task, with variations in dry friction at the start of the movements, backlash and deflection (considered here as a source of vibration of a servocontrolled system). Starting with a frequency of the order of 10 Hz, the duration of execution increases by about 20 per cent for 5 h 20 min. But if the frequency is set at 2 Hz the time is doubled. In parallel with this, the introduction of 8 cm of play extends the operating time by a factor of 1.5. Finally, the introduction of friction at 50 per cent of the maximal torque of the motor prolongs the time required for the operation by about 25 per cent. The figures are given here in a reduced form, but the study includes a multivariable analysis of their overall effect. This result is typical of the task studied, because tracking only involves estimation of the displacement and there is no interaction involving the manipulator in contact with an object.

On the same principle, Hill (1976I, II) adopted a far more mathematical approach in choosing the transmission delay between a movement in the master arm and the resulting movement in the slave arm as the main variable. In this he was effectively continuing the work of Sheridan and Ferrell (1963), but found very different results, which illustrates the importance of experimental conditions (Freedy and Weltam, 1972). The experiments were carried out on the Rancho Arm, which was used as the master arm (orthotic device), and another was used as the slave arm. The task was to take an object from above and transport it to a receptacle, some distance away. The only feedback signals were visual signals, and positional servocontrol carried out in this way has certain drawbacks (eg limited velocity, juddering of the slave arm).

When there is a transmission delay, the general strategy of the operation is known as *move and wait*, which means that a small displacement is carried out in the master arm, and then effect on the slave arm is observed, and so on.

Hill considers seven parameters which he believes to be pertinent and he attempts to find out which of them are related to each other, thus allowing only a few merit factors to be derived from the system. These values are the number of master movements (M) and slave movements (S), total energy used (E), total duration of the task (T), total duration of movement in the master (MT), MT/T and MT/M.

From this, he deduces, after studying the correlations, that the performance factors with importance in the delay are: M, T and MT/T (the work E is very closely correlated to the time).

If the execution time for the task without any delay is examined (see Kugath's example), it is of the order of 20 s. Here then are two teleoperation systems executing tasks with very little difference except in duration (and yet the second, and better, does not include any force feedback). Can one of these be selected on the basis of this single merit factor? Certainly not.

If we take a closer look, based on the differences concerning force, the tasks are similar at the moment at which the object is grasped, but essentially different when, for one the insertion requires great precision and force control and for the other it consists of throwing an object into a box, which requires no precision or force control.

Bejczy (1975, 1977) takes the execution time of the task as the only important property. There are four different tasks.

Task 1: moving from an imposed starting configuration to a rectangular block A, grasping it and placing it with the best possible alignment on an identical block B [see Figure 3.1.3.(2)].
Task 2: moving from an imposed starting configuration to a box, inside which a rectangular block is hidden. There are two possible cases: one side of the block may be open or the top of the block may be open.
Task 3: moving from a starting configuration and grasping an uneven pebble placed far away on a table.
Task 4: this is identical to task 2 but the rectangular block is replaced by a pebble.

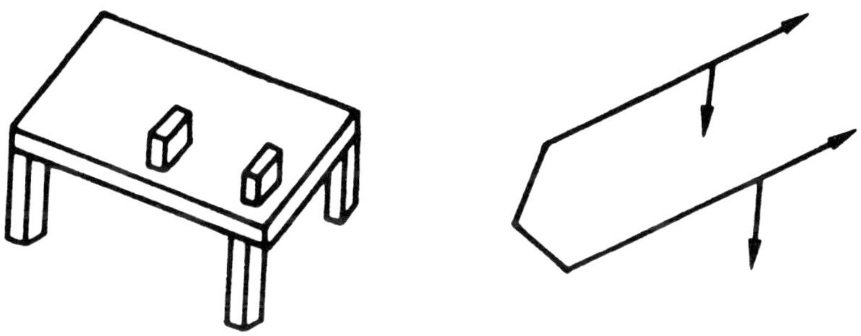

Figure 3.1.3.(2) *Bejczy's test tasks (JPL-USA):* **1**, *initial position of the gripper;* **2**, *stereo television camera;* **3**, *mono profile camera;* **4**, *proximity sensors directed forwards providing the operator with two distinct sounds (not stereo);* **5**, *proximity sensors directed towards the arms, providing different sounds from the sounds supplied by sensors 4*

The remote manipulation system is the one perfected at the JPL using the Ames arm under unilateral position control [Figure 1.4.3.(6)] and a gripper equipped with four proximity sensors [see Figure 3.1.3.(2)].

The variable to be characterized by measuring the duration of the tasks concerns the sensory feedback equipment in the system and in particular the variations according to the presence and location of the proximity sensors on the slave gripper, associated with distinct sound feedback. The author examines nine examples of figures for tasks 1 and 3:

visual data (mono + stereo) alone:
- (average time for task 1 is 24 s, for task 3 is 25 s);
- stereo vision plus two forward sensors (25, 24);
- stereo vision plus two bottom sensors (27, 25);
- stereo vision plus four sensors (31, 32);
- mono vision plus two forward sensors (22, 24);

– mono vision plus two bottom sensors (23, 21);
– mono vision with four sensors (30, 29);
– only mono vision (36, 39);
– only stereo vision (39, 38).

In the same way the following cases are recorded for tasks 2 and 4:

1. box with one open side and visible from two sides:
   – mono vision plus two forward sensors (29, 52*);
   – mono vision plus two sensors at bottom (41, 43*);
   – mono vision and four sensors (52, 61*);
2. identical to the previous case but one side of the box is not visible:
   – (28, 56*);
   – (44, 41*);
   – (57, 68*);
3. box with an open top (view from above):
   – stereo and mono vision with two forward sensors (52, 28*);
   – stereo vision plus two forward sensors (64, 72*);
   – mono vision with two forward sensors (92, 81*).

Note that the performance considered here is no longer just that of the teleoperation system itself, but that of the operator in the system, as a function of the improvements that can be made to the system. This concerns *differential performance*, and this relationship between the operator and the teleoperation system is considered in detail in Chapter 3.2.

The work of Jean Vertut at Saclay dating from 1973 has led to an attempt being made to compare different remote manipulators on the basis of four tasks believed to be pertinent (Vertut, 1973I, II, 1975). All these tests were directly viewed from a distance of about 1 m. A comparison was first attempted between the man-manipulator performance for different mechanical remote manipulator models (see Chapter 1.3), for a number of motorized manipulators under analytical control (see Chapter 1.4) and then for a manipulator with positional control by exoskeleton without force feedback [see Figure 1.4.3.(6)]. Figure 3.1.3.(3) shows that work using these manipulators (the execution time by hand for the first task of grasping, depositing and inserting being about 11 s for a return journey at 40 cm was taken as the reference) requires a duration time multiplied by a factor of between 1.5 for the lightest mechanical master-slave systems and 100 for the slowest motorized manipulators with key-pad control. For the assembly task involving several components, which took 6 s (5 s for dismantling), again taken as the reference, the time duration is increased by a factor of between 3 and 200 for a variety of manipulators. This work made two basic contributions to the understanding of the issue.

***This indicates that some of the tests failed.**

The first, mentioned above, allows a comparison to be made between the total time with the same operator in the loop and the same task, and illustrates the difference for execution of the same task using control modes with kinaesthetic feedback (situated between a factor of 1.5 and 2.5), and using positional control without force feedback (factor of 10). Kugath and Wilt (1973) established a factor of 1.5 (with good visibility) when comparing insertion with and without force feedback. The duration (88 s) is still surprising, as opposed to 25 s with the MA23 for two successive insertions. Note, however, that the manipulator without force feedback used in this experiment had exoskeleton control [Syntelman, Figure 1.4.3.(5)]. The new delay may be caused by the limited velocity of the Syntelman or by the difficulty in control relating to the exoskeleton itself.

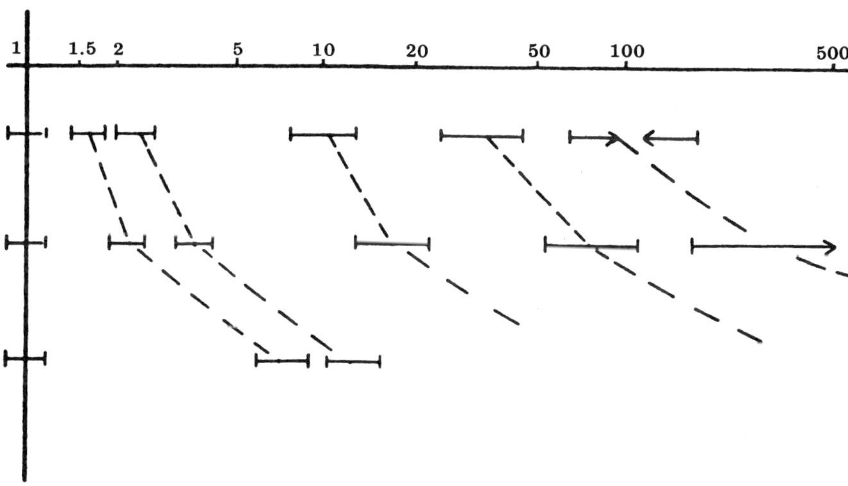

Figure 3.1.3.(3) *Time efficiency quotient for different remote manipulators (direct view, no delay) (Vertut, 1973):* **1**, *task of grasping and placing a pierced cube onto a vertical spike;* **2**, *assembly task involving insertion of a cylinder into a support, insertion of a rod into the unit, introduction of a pin (the orientation of the shaft must be correct to allow the pin to enter);* **3**, *extrapolation of a complex task (maintenance), involving a large transporter, multiple television cameras and delays in transfer of tools. Measurements taken during tests using an MA23 manipulator [see Figure 1.7.10.(1)] and no computer verify the results estimated in 1973.* **4**, *light mechanical master-slave manipulator (MA11 or the equivalent) or bilateral servo manipulator with low capacity and inertia;* **5**, *master-slave manipulators with a capacity of 10 to 20 kg (mechanical and servocontrolled, such as the M8, model M and MA23 manipulators);* **6**, *manipulators with positional servocontrol of extremities but lacking force feedback. Figures taken from tests using the MA23 manipulator without force feedback by Hill and Vertut.* **7**, *positional control by exoskeleton (Syntelman);* **8**; *coordinated velocity control;* **9**, *analytical control of velocity using six levers (PaR);* **10**, *analytical control at constant velocity by push-button (ACB)*

On the other hand the difference between the MA11 manipulator and the M8 manipulator lies only in the backlash, and, above all, in the inertia. It brings about a significant factor of prolongation in the time required, due this time to the reaction of the operator in not exerting sufficient force to overcome the inertia. Finally, the duration of the task with an analytical position controlled manipulator (by push-button) or with analytical position and velocity control [PaR manipulator in Figure 1.6.4.(3)] displays an effect due to its own velocity mixed with the effect of the load in the operation to combine the independent movements, without it being possible to distinguish the effects separately.

The other new aspect of this work is the comparison of the duration of the same task performed by hand by the same operator, which has led to the proposition of a time efficiency quotient, which was also considered at the same time by Hugatti (1973, p. 183). This work allowed an explanation of the difference between this coefficient of dexterity in the insertion task and the coefficient provided by the more complex assembly task with three parts. The explanation results in the gripper of the manipulator allowing the task to be performed using the same procedure as in manual execution but deprives the operator of a large amount of tactile sensory feedback to the hand and fingers. These factors will be taken as a reference to introduce another factor which appeared in other tasks in the same test process: a coefficient of strategy linked this time to the structural difference between the manipulator end effector and the human hand. Turning on a tap using a manipulator consists of a series of thirds of rotations. If the operator is required to turn on a tap, or unscrew something only using two fingers in the way a manipulator end effector does, then a time efficiency quotient similar to the previous one is found.

This work has been recommenced recently for complex tasks over long periods involving a single operator, a large transporter and several cameras. A coefficient of about 5 or 6 overall assuming direct vision has been established. The time efficiency quotient with television is about 8 or 9, and includes 30 per cent of the time in bilateral mode, 30 per cent in unilateral mode for the transporter and 40 per cent in unilateral mode for the cameras. Using automatic camera tracking the direct view time is approached.

3.1.3.5 FINDING A REFERENCE

The first striking detail noticed when watching the execution of a simple task involving a teleoperation system (eg grasping an object and joining it with another using a simple mechanical connection, like a peg in a hole with high tolerance) is the slowness or great difference in the duration of operation (from 1.5 to 88 s with good visibility) depending on the system used.

But it is important to ask the question 'slow relative to what?' Implicitly we make a comparison with three cases already encountered.

1. As the operator stays in control of the teleoperator, the system appears to carry out the task in the same way as a human, and executes tasks occurring in every day life, and the comparison with manual work is immediately made. (How can it take 88 s to push a peg into a hole when I can do it by hand in 2 s?).
2. This comparison cannot be made when the loads transported become too heavy to be lifted by hand. 88 s to insert a 40 kg cylinder into a ·300 mm diameter hole with play of 1 mm will appear acceptable and perhaps even quick taking into account the time that would be required on a building site to perform the same task with a crane.
3. The other comparison instinctively made is that of an automatic machine specializing in one task, which can perform the operation in a few tenths of a second, far more rapidly than by hand.

These three references, which only take the parameter of time into consideration, do not all have the same value. The third should be ignored straight away, because an automatic system excludes the permanent presence of an operator and has no versatility, since this is sacrificed to the speed and high levels of productivity (high-level robots which, as yet, do not exist in practice are excluded here).

The first two comparisons are valid because they are made in relation to two ways of execution which retain the invariant parameters of teleoperation: presence of an operator, variety of possible tasks and variety of ways of executing a task. It would therefore be legitimate to concentrate on two classes of reference for performance measurement:

1. The human reference if the tasks can be executed manually with or without tools used directly by the person involved, with an amplitude of displacement possible for man, as distinct from the arm movement alone corresponding to the work of a seated operator and that of movement of the order of 1 m corresponding to a standing operator.
2. Another reference can be used if the loads are too heavy to be moved manually. It is, therefore, the effective load and the size of the displacements that may separate the two main classes of teleoperation system, using this method. The fact of abandoning the mechanical master-slave manipulator, a tool which 'transports' the human hand on a unit scale of forces and displacement at a small fixed distance, in preference for teleoperation (which implies the use of remote servo manipulators and displacement of transporters and vehicles) reveals, as was the case when discussing classification (1-2), the two axes of the scale effect relative to the human manual reference: the scale of force multiplication on one hand, and the scale of space multiplication on the other.

Finally it is important not to forget the effect in the opposite direction, which leads to the hitherto unexplored areas of mini- and micromanipulation. Only research into human references appears to open the way to other work. In order to determine a bare handed work reference, two categories of questions must be answered:

1. To find highly significant standardized tasks or a way of breaking any task down on the basis of very simple primitives. This problem is difficult to solve because several ill-defined phenomena are involved (see Volume 1 of this series entitled *Modelling and Control*).
   (a) The level of complexity of a primitive must be determined.
   (b) Any task may be executed with an infinite number of combinations or arrangements of primitives.
   (c) There must be compatibility between the primitives chosen *a priori*, that is without taking account of the system which will execute them, and the primitives that can be executed by the manipulator.
   (d) Tremendous problems arise from the non-stationary or non-invariant quality of the primitives that can be executed by the system with respect to the work space when a performance is to be evaluated. For example, a displacement of 2 cm in one direction may be easy in one zone of the reachable space and very difficult in another. This is why, in all the work mentioned above, the tasks chosen always follow a heuristic choice with regard to their complexity or elementary status.
2. To find values to measure that are suitable for analysis in terms of merit factors. In this case also, although the time factor appears to be the governing one, it implies that the task as a whole is correctly executed right to the end. But many tasks (eg trajectory tracking) cannot be measured by a single time factor. The human reference is justified all the more when a teleoperation system is to be rendered 'transparent', that is that the operator has the impression of working directly with his hands.

For instance, consider Peruchon's method (1975). He chose the following tasks:

- specifying an imposed rate between two points (the distance between which was a parameter between 3 and 70 cm) situated in a plane;
- tracking plane trajectories (squares, circles etc);
- assembling two cylinders.

He measured the time and precision with and without visual control after a period of training. In a second attempt he considered assembly by trying to measure the force involved. Unfortunately, the formalization of performance factors appears not to have been achieved yet, and it is clearly not going to be easy (Peruchon, 1983).

3.1.3.6 OTHER QUANTITATIVE FACTORS

Although they are intended for manipulators used as robots, the performance criteria suggested by Konstantinov and Genova (1981) should still be taken into consideration for remote manipulation. They are also divided into four classes, and summarize part of the work of Ozaki *et al.* (1977) and Warnecke and Brodbeck (1977).

Criteria for characterization of structures

1. Degree of anthropomorphism (geometric): this is the ratio of the number of rotation (R) DOF to the total number of DOF. The other DOF are translations (P);
2. Succession of R and P DOF;
3. Dislocation of successive axes (normal, parallel);
4. Distances between these axes;
5. Possibilities of development of the structure in a plane and along a straight line;
6. Total kinematic efficiency (displacement of a point on the end effector in uncoupled manner along the three Cartesian axes).

This first class of criterion concerns an analysis of the kinematics of the mechanism in the form of coefficients, and relates to the description given by Roth (1976). See Volume 1 of this series entitled *Modelling and Control.*

Criteria for accessible surfaces in the work space

This is an analysis of the reachable volume (see also Vertut, 1973; Lukishov and Milosardin, 1973).

Kinematic criteria

Velocity coefficients:

in translation: $K_v = V_{max}/V_{average}$
in rotation: $K_\omega = \omega_{max}/\omega_{average}$

Acceleration coefficients:

in translation: $K_a = a_{max}L/V^2_{average}$
in rotation: $K_\epsilon = \epsilon_{max}/\omega^2_{average}$

Rapid action coefficient:

$$t = \int_{q_{s_1}}^{q_{s_2}} dq_s/\dot{q}_s$$

q_s is a generalized variable.

Force and energy criteria

Maximum torque or force R felt by an articulation with average torque:

$$K_R = |R_{max}| / R_{average}$$

Peak energy over average energy:

$$K_T = |T_{max}| / T_{average}$$

This theoretical analysis is also complex and only provides the possibility of a single mechanism in this certainly interesting, but sometimes difficult, synthesized form. Can it really be that the degree of anthropomorphism is measured by the number of rotations?

Morecki adopts a more complete concept of the degree of anthropomorphism, by adding to the geometrical approach which concerns only the number of independent movements, the number of circuits for control and sensing. This attributes a poor manipulator, which is apparently anthropomorphic with a coefficient of 0.12 for the motricity and 0.23 for function (Morecki and Eckiel, 1978).

Ananiev and Nakhapetjan (1981) devised experimental tests in which he measures the precision of a given robot by tracking the end effector during imposed tasks and evaluating the expression:

$$Q = \frac{1}{t_2 - t_1}\left\{\int_{t_1}^{t_2} \left|\Delta X_T(t)\right| dt + \int_{t_1}^{t_2} \left|\Delta Y_T(t)\right| dt\right\}$$

in which t_1 and t_2 are the initial and final times of trajectory tracking; ΔX_T and ΔY_T are the deviations with respect to the trajectory (taken tangent and at right angles to a point in a fixed plane).

Conclusions

Figure 3.1.3.(4) shows approaches to performance evaluation of teleoperation systems. It is clear that we are still far from achieving perfect unity. Any teleoperation system is so complex that any specialist, in any field, could find a different theme to consider. Faced with a complex system there seems to be no clear simple way of placing it *a priori* according to a potential mode of use.

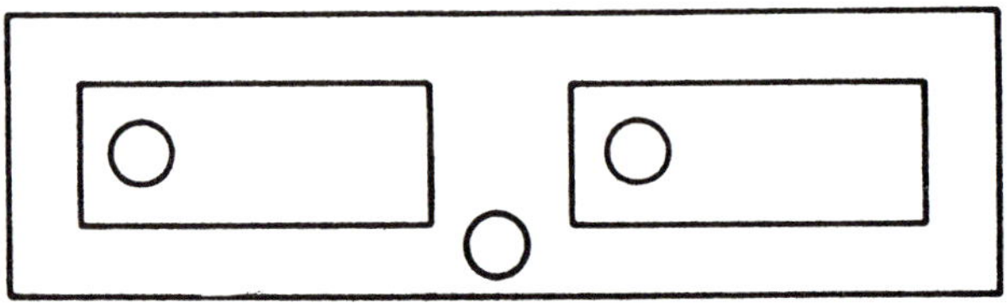

Figure 3.1.3.(4) *Types of measurement taken for performance evaluation*

Chapter 3.2

The human operator in the teleoperation system

This chapter was written by Nicole Fiori and James Richardson of the National Laboratory for the Study of the Physiology of Movement (ERA CNRS No 956), University of Paris XI, 91405 Orsay.

3.2.1 Introduction

This chapter is based on the contribution of the man-machine interface to the development and evaluation of remote manipulators and teleoperators.

The use of new technological constituents now allows the design of remote manipulators which are better suited to the operator's needs. In the design of new systems the technical choices should be based on the man-machine interface considerations so as to accommodate the human operator in the best possible way.

In parallel there has been an increase in the use of master-slave manipulators in the nuclear industries, and ergonomics can help in the design of workstations, taking into account the physiological and psychophysiological demands of the operator.

In a standard complex teleoperation system, there is a workstation which unites the conditions of the remote manipulator workstation and the control desk for all the ancillary parts including, in particular, the unit for feedback of visual, auditory and other signals. The operator is separated from the task by a physical distance for remote work, and by a shielding when the work is to be carried out in a hostile situation.

In practice, the hostile environment is usually also situated a distance away (motorized manipulators). Separating the operator from the work area reduces the reality of the work, and this is demonstrated by the reduction in data available from the task and the environment, but also from the consequences of his actions, and this causes a reduction in his capacity to perform a task.

Interaction between the operator and task consists of a transfer of data between the operator, the teleoperator, the remote manipulator(s) and the peripherals (Tesar and Lipkin, 1978I, II). The ideal remote manipulator should allow the operator to work at a distance with equal or even greater efficiency than when working bare handed. Many authors (Corker *et al.*, 1981; Flatau *et al.*, 1972; Charles and Vertut, 1974) feel that this cannot be achieved without telesymbiosis, telepresence and system transparency.

These terms describe the impression that the operator is working directly on the task rather than at a distance, and it is difficult to quantify the variables that may contribute to this impression. To develop telesymbiosis in a man-manipulator system, it is necessary to know the role of the operator in the man-manipulator-task loop. Ergonomics can provide data on the physiology and psychophysiology of the human operator, and evaluations of equipment and facilities used in teleoperation. Man-manipulator interaction is added to the role of the operator in guiding a vehicle or transporter, which is less complex in itself, but the overall complexity results from the combination of all these operations, and may require several operators.

The teleoperator and human operator carrying out the task of remote control and/or remote manipulation may be considered as a man-machine-environment system, in which the operator is an integral part. Such a system is made up of many parts which have a common object. The system outputs depend not only on the characteristics of each element but also on their mutual interaction (De Greene, 1970; De Montmollin, 1967). The system is represented in Figure 3.2.1.(1).

Interaction between the operator and remote manipulator can be analysed in the framework defined by the man-machine system, which allows a systematic examination to be made of the physiological and psychophysiological demands on the operator in developing an approach or carrying out a task. The design of the interface allowing the operator to communicate with the machine should take into account his abilities, and the success of the type of system shown in Figure 3.2.1.(1) depends on matching the properties of the remote manipulator with human abilities.

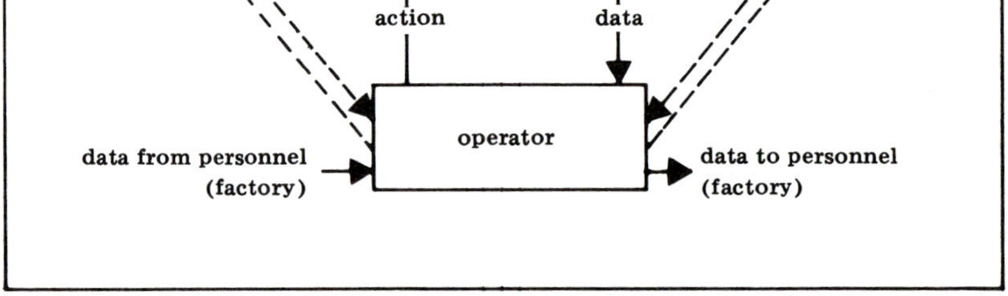

Figure 3.2.1.(1) *The man-machine-environment system*

In evaluating the ergonomics of remote manipulators, it is difficult to predict all possible circumstances that may arise during actual work. This is why the workstation and operator work load must be subject to ergonomic consideration.

The aim is to relate the design of the equipment to the abilities of the human operator so that the work is at a level that will not endanger the health of the operator. Overloading (whether physical or mental) the operator will have an adverse effect on the performance of the whole system.

Evaluation of the work load should be made in the real work situation, so that all contributing factors can be taken into consideration. It is a complex function of variables derived from the design of the equipment, the physical environment, the task and the organization of work. The effects of the work load will vary according to the characteristics of the operator (eg weight, height, experience, level of training).

Such evaluation cannot be considered valid unless applied to the remote manipulator in its normal position, used in the normal way, with the operators taken into account. The aim is to find possible examples of cases where the system is not adequate for the intended task, and there are possible inadequacies between the operator's abilities and the imposed conditions.

For this a description of the real constraints must be made with respect to the operator, and the importance of this being based on real working conditions has been shown by Wisner (1981). The study of the behaviour of the operator in response to the different constraints imposed by the task is only useful if the demands of the system are identified.

Evaluation of the work load is based on a comparison between the data on the constraints of real work and the physical and mental cost to the operator. There are different methods of analysis of work which allow identification of the work demands, namely:

1. physical demands (static, dynamic effort);
2. environmental demands (noise, lighting, radiation, movement);
3. sensory demands (hearing, sight);
4. sensori-motor demands (dexterity, movement program etc);
5. cognitive demands (memory, decision making).

The analysis of the operator's behaviour in response to the task concerns:

1. general physical ability and sensori-motor ability;
2. ability to gather and assimilate data;
3. cognitive and decision making ability [see Figure 2.1.6.(1)].

The methods of analysis in this chapter are taken from the study made by Leplat and Cuny (1977).

3.2.2 The man-manipulator interface

Communication between the man and the teleoperator is shown in Figure 3.2.2.(1).

There are multiple inputs concerning the senses, and the outputs from man to machine pass through the operator's movements. The operator must detect, identify and interpret data, which involves both short- and long-term memory and the decision making ability of the operator (see Chapter 2.1).

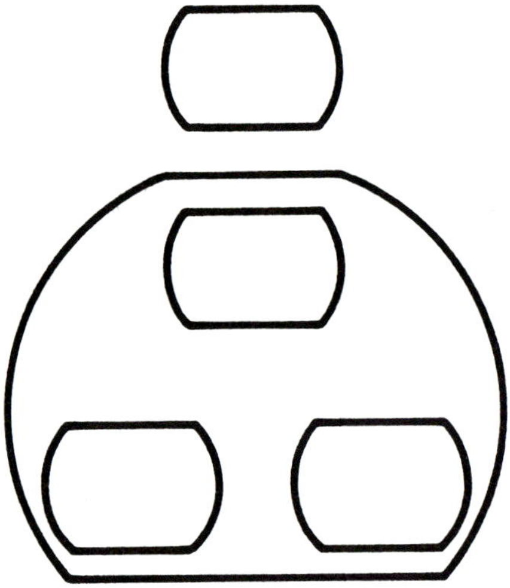

Figure 3.2.2.(1) *The man-machine interface*

The inputs may be direct or indirect, on line or deferred. Direct input implies that all the data are presented to the operator via, for example, windows, monitors, loudspeakers or displays. Man-machine communication will be considered in the context of the sensory and motor aspects.

3.2.2.1 SENSORY ASPECTS

Sight

The organ of sight in man is the eye, which functions as a complex focusing and directional system. All visual signals are processed by the retina, the thalamus and finally the visual cortex. Binocular vision, which allows perception of a single image, results from processing in the cortex which synthesizes the two images coming from the two eyes.

It allows, in particular, an object moving through the depth of the visual field to be followed, and contributes to the perception of relief. In binocular vision, without movement of the eyes, the visual field is 180° horizontally and 120° vertically. The conditions of vision in teleoperation are generally severely disturbed.

The field of vision is very much reduced. Using direct viewing only, there are frequently parts of the remote manipulator that obstruct part of the field of vision and the distance of observation is greater than that in normal work, which causes a reduction in three-dimensional perception and resolution.

In nuclear applications, viewing through a protective window poses a number of problems:

1. the field of vision is reduced because of the dimensions of the lead glass panes, which for technological reasons are limited;
2. the vision of objects creates an impression of closeness of the order of half the thickness, associated with a deformation effect which is greater when vision is oblique;
3. the sharpness of vision is reduced because lighting is often dim as a result of absorption by the glass and the lack of contrast between the different objects manipulated (which are often metal and uncoloured) and the interior of the enclosure, lined with stainless steel.

Paint cannot be used to distinguish objects in corrosive situations with high levels of radiation, and the thickness of lead glass causes a loss of perception of blue because of the intense yellow colour of the window, added to the fact that the lighting is not monochromatic (sodium lamps). There are many reflections within the enclosure, which contains projectors of several kilowatts, and on the operator side, since it is often impossible to protect it from the lighting behind the operator, for practical reasons relating to the work space.

All these factors can be applied to a manned submarine, with the difference being that the loss of colour is experienced in the yellow range because of the distance between the operator and the underwater application. For work at great depths, the limitations of the size of the porthole are an added disadvantage, and it is rare to be able to use it without some other visual aid.

When the operator works from a space capsule with a direct view of space, there are added problems because of the extreme contrast in lighting and the lack of horizon as a point of reference.

The use of television cameras can add to the problems because of screen flickering, which is characteristic of cathode ray tubes, and the angle of vision depending on the camera position, which is usually more or less fixed, and may be different to that of the operator in conditions of direct viewing. In addition, stereoscopic effect is lost, the lack of contrast may be made worse in black and white and the field of vision is isolated from the work space etc.

Systems equipped with two or more cameras require adaptation by the operator because, unlike under normal visual conditions, these systems involve the acquisition of data from different sources. Stereoscopic systems equipped with two cameras and only one monitor, which attempt to reproduce the human visual system, can only be used correctly when the operator wears special glasses, which may be considered a disadvantage.

The use miniature monitors fitted inside a helmet with sensors for orientation of the head (Charles, 1976; Foote and Dompe, 1978) may be even more unpleasant for the operator, and restricts the use of the visual information to the operator alone. It does, however, provide orthoscopic vision for a value of the focal length of the camera lenses.

In conclusion, the visual aspects of the man-teleoperator interface modify the normal condition of vision to a considerable degree, and in order to detect, identify and interpret the visual information, the operator cannot simply rely on his normal reference system.

Hearing

Sound waves reach man through the ear, which is made up of three parts: the outer, middle and inner ear. Perception of sound depends on the frequency and intensity of the stimulus.

The total field of perceivable frequencies ranges from 16 Hz to 20,000 Hz. The threshold for audibility of a sound, or the minimum intensity depends on its frequency; the minimum falls between 1,000 and 6,000 Hz and increases for higher and lower frequencies.

High intensity auditory stimuli, and prolonged exposure to noise, may cause temporary increase in this threshold, or momentary deafness.

When two auditory stimuli of similar frequencies are emitted, the operator either detects a single sound, which is a combination of the two, or the sound with the greater intensity, constituting a mask effect.

Finally, a source of sound can only be located if both ears function normally, so as to allow determintion of the difference in time of travel and relative intensity of the sound in either ear.

In remote manipulation, the operator is subject to auditory stimuli produced in the master environment. For example, in a nuclear facility, noise from the ventilation system, apparatus for atmospheric testing to control contamination, remote manipulators in action, apparatus at the workstation etc may be heard. These noises may constitute an auditory background with an intensity of 60 to 80 dB at frequencies between 65 and 4,000 Hz. To this can be added other possible noises such as conversation by interphone, possible alarm signals etc.

These conditions in industry will improve as a result of ergonomics, and future facilities should free the operator from this background of noise in the work environment. In underwater work, manned submarines are relatively quiet so that this problem does not arise.

Increasing importance is being attached to high quality stereo sound transmission for teleoperation, and this is becoming more feasible with the use of bilateral servo manipulators which can be used at a greater distance, but direct viewing through windows is lost.

The above facts indicate that the operator's acuteness of hearing may deteriorate because of the background noise (mask effect or fatigue) or the temporary increases in the threshold of audibility, and this may lead to difficulties in hearing the sounds relevant to the task in hand.

Touch and kinaesthesis

The displacements, velocities and accelerations in different parts of the body imply a diverse system of sensors throughout the human body. The first of these sensors are the mechanoreceptors, activated by touch, pressure on the tissues, the positions and movements of the joints. In the skin, some of these receptors are sensitive to pressure as a function of time, and others provide information on the intensity and velocity of application of the stimuli. In the articulations they are sensitive either to the position of the articulation and the detection of its movement, or to the velocity of acceleration of movement. In the tendons, the receptors are sensitive to muscular tension, whether resulting from contraction or extension of the muscular fibres, whereas in the muscles the neuro-muscular fibres, situated in the fleshy parts of the muscles, are sensitive to the length of the muscles considered.

To these mechanoreceptors is added the effect of the vestibular and visual sensors (Berthoz, 1981). The vestibular sensors are found inside the inner ear and are sensitive either to angular acceleration and thus to rotation, or to linear acceleration in the horizontal and vertical plane of the head and to acceleration due to gravity. This allows the position of the head to be detected.

The visual information can supplement the lack of vestibular data when the movements are made at constant velocity, since the vestibular sensors are only sensitive to acceleration, and are also unable to distinguish between a linear acceleration and the inclining of the head. Finally, the visual information allows perception of movements in the scene occupying all or part of the visual field.

All this information, provided by different parts of the central nervous system, is processed and integrated, and provides the operator with a degree of awareness of the position of his limbs, which is known as kinaesthesis (Albe-Fessard and Tyc-Dumont, 1976), but also allows impacts and vibrations etc to be detected.

Information of this type is clearly different depending on whether manipulation is carried out bare handed or with a remote manipulator. When an object is moved manually, the acquisition and processing of information involves the normal system of references contained in the long-term memory. When the same object is displaced using a bilateral master-slave manipulator the information is quite different: the tactile information felt in the hand originates from the handle and not the object, and making the correspondence between the kinaesthetic information and the visual information on the position of the object is more difficult, because the operator has to use information relating to the vibrations and friction transmitted (but considerably filtered) by the remote manipulator and the handle.

There is still little information available on the problems in control stations on board ships, but it is known that the control of a remotely controlled vehicle (helicopter, submarine with or without manipulator) can be considerably affected by the movement of the ship. The introduction of bilateral servo manipulators into this field will involve the solution to the problem of interference in kinaesthetic control by the kinaesthetic perception of the ship and may involve stabilization of the control station on board.

3.2.2.2 MOTOR ASPECTS

Professional tasks, even the most simple, always seem complex because of the demands involving force, velocity and precision (Bouisset, 1981). The complexity of movement is no less important and affects the various parts of the body in different ways. The performance of these movements must be compatible with maintaining balance, implying the adoption of appropriate postures. Any motor act not only requires the cooperation of the muscles directly involved but also those for the necessary posture to stabilize the movements.

The displacement of part of the body results from the successive or concomitant movements of the joints concerned. Generally, a distinction is made between the movements transporting the limbs, involving the joints and proximal muscles, in the shoulder and elbow for the upper arm, and the small movements such as palpation, which is connected to the action of the distal joints and muscles, in the wrist, hand and fingers, for the lower arm.

Any joint movement is based on the contraction of a specific group of muscles opposing movement. For example, the flexion of the elbow is ensured by simultaneous contraction of the biceps, brachio-radial, brachial-anterior and the pronator ring. The muscles opposing this movement, which are also antagonistic to the opposing movement, contract to a greater or lesser extent, both setting the joint in motion and stopping the movement.

Muscular contraction is caused by electrical activity in the muscles, or electromyographic (EMG) activity, which depends on the velocity of the movement. In rapid movements, there is successive contraction in the agonistic and antagonistic muscles, in slow movements braking is ensured by the passive forces, such as inertia and the visco-elasticity of the joints, and the antagonistic muscles are not involved. These results relate to simple movements (Lestienne, 1974), but introduce the concept of planes of muscular cooperation. Such planes are vital to the coordinated movement of joints. The concept of developed force should be further explained. A distinction must be made between:

1. the muscular force itself, produced in the tendon of a muscle, and proportional to the length of the muscle, depending on the velocity of contraction of the muscular fibres;
2. the external muscular force resulting from the contraction of various muscle groups, which depends on:
   - the relative position of the segments of the limb exerting the force, or the articular angles for which the muscular force is largest;
   - the nature of the force in terms of thrust, traction, raising, lowering and rotation;
   - the posture which is involved in maintaining balance and in the reactional forces which may be developed in the supports.

The introduction of a mechanical master-slave manipulator and the standing work involved in the execution of a task interferes with the motricity, as briefly described above. Small movements of the fingers are impossible when the hand is expected to hold a control lever, and the only possible movements are, for example, those that involve pressing push-buttons or switches. The postural modifications for application of a maximal force are limited by the organization of the workstation, but mechanical phenomena such as extension of the leverage on the slave arm (Z indexing motions, see Section 1.3.2) must be compensated for by the adoption of specific postures, such as support of certain parts of the body on the master arm.

Apart from handling weights of up to 300 g, which can be carried out using one hand, and allows two manipulators to be controlled at once, most operations with mechanical master-slave manipulators involve force levels requiring one manipulator to be controlled with both hands. This is improved with the use of bilateral servo manipulators on large transporters, in which the operator only has to perform small movements, with force reduction in a ratio of Y4 to Y8.

3.2.3 Demands of the remote manipulation task

Manipulation tasks often involve the transport of object and/or position with respect to an oriented target situated in the space. Transport involves displacement characterized by direction and amplitude. Positioning may constitute a constraint depending on whether the object is simply to reach a point in the space or to be precisely adjusted with respect to the target, with a reduced number of DOF, as is the case for insertion tasks with objects of particular shape. These operations must all be carried out with a certain degree of force and precision, which has direct consequences on the mobility of the segments and the use of the various muscles in the operator's body.

The mechanical master-slave manipulators described in Chapter 1.3 almost always reproduce the position and force with a scale of one, but introduce a parasite mass variable for the position and direction but always of the same order as the handled mass. This mass added to the effect of backlash, which exists in certain movements, constitutes a constraint on the spontaneous movements of the operator. This *mass feedback* due to the inertia of the system provides the operator with an opportunity to establish dynamic control, and causes fatigue when the manipulator is unloaded, and slows down the movements spontaneously. This retardation explains the time efficiency quotients of 1.5 for a light manipulator without backlash and 2.5 for a telescopic manipulator, in which the inertia is tripled. The mass feedback should be constant and memorized by the operator during the training phase, as a function of the position of his own arm and the manipulator. In this way the operator can minimize the volume in which he moves his hand, by using a system for electrical indexing in most manipulators. Control using push-buttons on the control lever takes time, and the operator will choose between the fatigue of this control and that of the load, but in either case the operation will be slower. The inevitable flexion depending on the direction, which forces the operator to handle the object while compensating for the dynamic effects of this, is also added to the above considerations.

Using bilateral servo manipulators it is possible to manipulate loads outside the scope of direct mechanical remote manipulation, and the ratio of force feedback is generally 1/4, which also reduces fatigue. The operator is also aware of reduced inertia, and can profit from this. This choice of ratio on the force feedback, and of other characteristics, such as friction and flexibility, has an ergonomic use. For manipulators with larger capacities, such as the Andromat (CSEE, France), initially developed by General Electric USA as the Man Mate, the scale effect may reach 10 for displacement and 1/100 for force.

In current remote manipulators the lever is not a replica of the object held in the gripper, and the operator holds this object as if he were carrying it with a gripper at a distance. This requires, in addition to the grasping force, a suitable torque to support the load at a certain distance.

This can lead to stiffness in the hand and considerable local muscle fatigue, which is generally manifested by a weakening of the force applied and thus loosening of the gripper and loss of the object held. Locking or electric grasping in the gripper partly reduces this risk in that it can prevent the object from falling, but the operator must continue to hold the lever, if only to direct the master arm, and fatigue is still a problem, leading eventually to a real loss in dexterity.

Figure 3.2.3.(1) (Vertut, 1975) shows the results of work carried out at Saclay, comparing the MA11 mechanical manipulator with different types of handle providing varying degrees of interaction between the

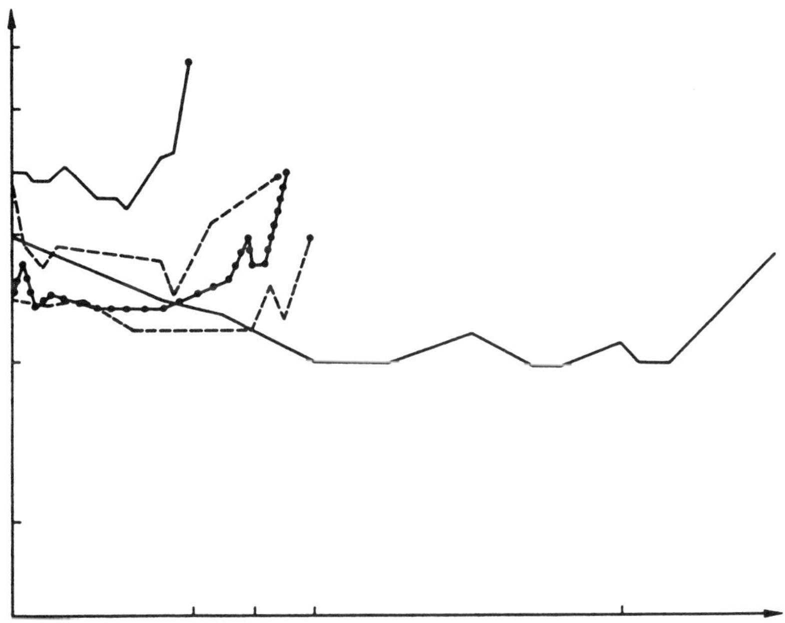

Figure 3.2.3.(1) *Fatigue test with different types of handle: the test consisted of putting up two rows of three lead bricks each weighing 5.5 kg, then reassembling them 50 cm away. The diagram shows the time of cycle T in seconds according to the number of cycles carried out by the operator during the test.* **1**, *time of manual cycle 17 s;* **2**, *number of cycles resulting in 1 tonne of bricks being handled;* **3**, *standard ambidextrous handle controlling gripper closing (CRL, USA) [see Figure 1.3.5.(4)(a)]. The use of this system causes a loss of dexterity (⩾3.8) and soon causes fatigue (1 tonne).* **4**, *pistol grip handle by Wälishmiller improved tests: 1.5 tonnes handled in 48 min;* **5**, *first PSM model [see Figure 1.3.5.(4)(b) CEA-La Calhène]. The ease of prehension results in an improved cycle time, and fatigue stopped the test after 1.5 tonnes had been handled in 45 min;* **6**, *improved model for fatigue (1.7 tonnes)*
Test with MA11 mechanical manipulator
Test with MA22 bilateral servo manipulator with 1/3 force feedback **7**, *handle 5 on the MA22, division of grasping and supporting force, fatigue tolerated up to 4.2 tonnes, or with 1/3 force and a duration 2.3 times longer*

grasping force and the torque applied in grasping the handle grip. Tests were carried out to establish the difference in fatigue when moving lead blocks each weighing 5 kg to build a wall, and then moving the wall block by block a certain distance away. The time taken at first reduced, and then increased, with fatigue until the operator stopped of his own accord. With both hands the operator could move up to 1 tonne a distance of 50 cm before stopping, whereas an improved handle allowed him to move nearly 2 tonnes. The same test was then carried out with the MA22 servo manipulator with 1/3 force feedback, and the same operator was able to perform nearly three times as much work before stopping. The division of force by a given factor would appear to have a more or less linear effect on the fatigue of the operator.

Another improvement in manipulator handles consists of using computers to transfer the centre of gravity of the load to the centre of the operator's hand, and cancelling out the torque required to support the weight, which is far less tiring. A third method involves cancelling out the weight (and not the mass) using the computer, which is a considerable improvement from the operator's point of view, and has been used in the ARA project at Saclay since 1983. The ergonomics have not yet been completed (Acquaviva-Lescoat, 1983).

Finally, one of the demands of remote manipulation is to take into account transmission delays which exist between the reproduction of movements in the master arm by the slave arm, in the different types of feedback from the work area to the operator, caused by problems of transmission over long distances and the time required for decoding. The sum of these delays results in feedback, which does not relate to the result of the operator's action, and leads to deterioration in performance (Ferrell, 1965, 1966; Hill, 1976I, II; Hill and Salisbury, 1977); beyond delays of 1 s, the performance is only adequate for open-loop tasks, where the operator does not use kinaesthetic feedback.

Recent work at Saclay by Micaelli (1982) should allow experiments to be carried out with good delayed force feedback.

3.2.4 Strategy for task execution

During execution of a task, sensory activity, activity of the nervous system and motor activity are closely linked.

Sensory activity consists of acquiring data on the environment, the nature and position of any objects and the situation of the operator's body itself. These signals are decoded, interpreted and related to memorized information, resulting in a decision being made on the appropriate strategy for task execution. This allows the operator to decide subconsciously on the nature and characteristics of muscular intervention required.

Many authors considering any finalized action envisage transport of the limb and precise adjustment to the target. The movement of the limb depends only on continuous feedback of kinaesthetic signals from the musculo-tendon receptors, whereas the final adjustment involves the discontinuous acquisition of signals from visual, articular and cutaneous receptors, so as to reduce the error between the position of the hand and that of the target.

Generally speaking, a distinction is made between continuous applications and point-to-point applications. The former requires that data are taken at all times allowing adjustment of the motor action, whereas the latter does not require such adjustments because the movement consists of a single action, such as pressing a button, or the task is formed of a series of isolated actions. In this case the task will be based on a motor program in which feedback is not required (Keele, 1968; Schmidt *et al.*, 1979).

Training may consist of implanting such motor programs, resulting in the confrontation of data from the space in which the movements take place and the sensori-motor data stored during the preceding movements (Paillard, 1972). Initially fragmentary and dependent on internal sensors, the movement becomes continuous and relatively independent of the messages relating to its execution (Massion, 1978). So training can allow the operator to move from a strategy based on continuous adjustment (with feedback) to one in the form of discontinuous adjustment, or a strategy breaking the task down into the transport of the object (without feedback) with precise adjustment to the target (with feedback).

In studying training of a simple targeting task carried out bare handed, Fiori-Savary and Richardson (1979) showed that training leads to an increase in the velocity of execution, and therefore a reduction in duration of terminal adjustment. The process continued as if, during training, the sequences of movements executed with dependence on a motor program, without feedback, became relatively longer (Schmidt, 1975). The reduction in the amount of data acquired during execution of the movement results in an overall reduction in the duration.

Many studies attempting to analyse the respective roles of direction, amplitude and precision of the action have concluded that direction is treated first, and gives rise to a motor program for transport. The duration of execution is a function of amplitude and precision, and Fitts (1954) proposed an index of difficulty (Id) for changing direction based on precision relative to, for example, the ratio between double the amplitude of displacement and the size, or tolerance, of the target. This author considers that the execution time is linked to this index by a logarithmic function. Note that Fitts law was established for unloaded manipulators running at maximum speed. Apart from the fact that teleoperation work generally consists of transporting loads, manipulator inertia, friction and backlash prevent it from operating at full speed. The work of Fitts is nonetheless valid, particularly when he considers the amount of data acquired and processed as constituted by the Id. The increase of this quantity would increase execution time.

The operator must always adopt strategies of a very different type to those he would use for natural movements, particularly when training can allow natural movements to be freed from feedback and be executed at least partly in open loop, and the hypothesis could be advanced that during remote manipulation work, movements remain dependent on the feedback of signals, in particular when the strategy adopted is that of closed-loop execution or in the form of *move and wait.*

Ferrell (1965, 1966) showed that when a transmission delay of 1 s is introduced into the force feedback from the master to the slave arms, the move and wait strategy is adopted spontaneously by the operators. When there is force feedback but no visual feedback, this strategy is preferable for the correction of the gripper position with respect to the object. For Hill (1976I, II), working without force feedback, this strategy consists of a series of open-loop movements followed by pauses to wait for the feedback of signals. It is strictly applied for transmission delays of 1 to 3 s; for delays of 0 to 1 s a closed-loop move and wait strategy is adopted; for delays of more than 3 s, complex movements, although unintentional are observed during the pauses, which can be explained by the fact that the subject cannot stay still for so long.

When the move and wait strategy is well established, the duration of the tasks can be predicted from the duration of the movements carried out in open loop and the number of these movements depending directly on the Id of the task (Ferrell, 1965; Hill, 1976I, II).

This proves that in every case the task is made up of movements and durations of movements which are added together, introducing the problem of *time and movement.*

The study of action sequences requires a knowledge of the characteristics of the task, but is helped, in the case of a complex task, by breaking the task down into successive elementary tasks (subtasks). The method used to solve this problem of time and movement is based on two principles:

1. manual tasks can be broken down according to the fundamental hand movements (Gilbreth *et al.*, 1924);
2. the total time taken to carry out the task is the sum of the elementary execution times for the basic movements which make up the task.

Leplat (1966) felt that this addition of execution times for elements of a task was only possible if the elements themselves were independent. He concluded that the elements should relate to the type of task and the method used.

Faverge (1952) and Ombredane and Faverge (1955) raised an idea that the study of time and movement could be based on an arbitrary breakdown of structures within the task time which would not be added together unless successive actions overlapped.

Davidson (see Leplat, 1966) noted that the analytical breakdown may be valid for beginners, but for experienced operators their work would be better defined in terms of 'groups of elements the units of which are generally in correlation'. This is partly in accordance with De Montpellier (1964), where it is stated that 'training consists of a systematic modification in behaviour, in cases of repetition of a single situation. . . leading to an improvement in performance', and Leplat (1970) who says that during manual training, the increase in speed is based above all on the modification of the mode of executing a task; in particular the manipulation time reduces by a greater proportion than the time taken for displacement.

Acquaviva-Lescoat and Fiori (Acquaviva-Lescoat and Fiori, 1982; Acquaviva-Lescoat, 1983) considered the possibility of breaking down a remote manipulation task into separate elements. The test task, which involves transporting an object and inserting it into and out of two shafts 40 cm apart, is derived from Vertut's first test (1973) and was carried out using the MA11 manipulator.

The test is broken down into the following phases: removing the object from the first shaft, moving it to the next shaft, positioning it with respect to the second shaft, inserting the object, then the reverse operation is carried out ending at the first shaft. The object can weigh 0.5, 2 or 4 kg. The operators were familiar with the manipulator but they were not very experienced operators. The results were as follows:

1. Positioning took longest (average of 2 s) and determined the total length of the task (average 8 s).
2. Increasing the weight of the object from 0.5 to 4 kg was manifested by a systematic and considerable lengthening of each phase. It was more marked during transport and positioning, and these times increased from 1.1 to 1.6 and 1.5 to 2.5 s respectively. The total time taken for the task increased from 6 to 10 s, whereas, for example, the length of the insertion phase increased from 0.5 to 0.6 s.
3. No significant correlation was observed between the lengths of the different phases.

These results have been taken to suggest the independence of the phases making up the task, and therefore the possibility of developing time norms for remote manipulation tasks taking into account the weight of the object involved.

The operators, it must be remembered, were not specially trained, but the continuity of electrical activity in certain muscles throughout the task and the existence of a plane of cooperation between the muscles have led various authors to conclude that the phases are not independent of each other in their physiological expression. In conclusion, it would appear that time norms in remote manipulation can only ever be used with caution and always with control of the physiological type.

3.2.5 The design of the workstation

The ergonomic design of the workstation is based on an ergonomic analysis of the man-manipulator unit. The activities involved in the execution of a task determine connection between the operator and his work, as in, for example, the manipulation of controls, on a physical level, and the acquisition of data whether visual, auditory or tactile signals, on the perceptive level. These determine the posture of the operator, and the ergonomic analysis should consist of establishing the existence and type of the relationship between the operator and the task. The tremendous variety of different situations makes the individual study of each case vital, but a certain number of standard procedures can be applied, based on the general ability of the operator.

A certain amount of anthropometric data is required first of all. These define the limited possibilities on the basis of which the ease values, which only represent a small fraction of the whole, can be fixed. The size of the workstation should depend on the variability of an operator, apart from the various anthropometric measurements; there is no such thing as a standard person. A percentile scale should be drawn up allowing the variation dispersion of individual data to be taken into account: 90 per cent of the subjects are contained between the percentiles 5 and 95. Note that although the correlation coefficients between the measurements of height and length of the upper limbs are high, the same is not true between measurements such as the height of the trunk and the different segments of the limbs, or between the forces of the different muscles. The multiplication of the demands of a workstation will rapidly exclude a good number of operators unless adjustment is possible. In any case, this illustrates the need to know the requirements of a task precisely to allow determination of the anthropometric measurements.

For example, when a mechanical master-slave manipulator is used and the work area is viewed through a window, it is vital that the height of the window is compatible with the height of the plane of vision of the operators. If the plane of vision is to be the same for all the operators, and the window is fixed, the workstation must include an adjustable platform, but this may present problems for the height of the upper part of the body, particularly for the articular angles relative to the master arm. These conflicting aspects of the workstation cannot generally be resolved except by choosing between the two, based on an ergonomic analysis of the tasks required and their demands.

The problem is solved when manipulators with indexing X motions are used (see Section 1.3.2). Experience has shown that the operator cannot often find the area of ease unaided, and automated systems would be useful here, within the framework of a computer-aided system, also applicable to control stations for bilateral servo teleoperators.

Anthropometric measurements have been studied in detail by a number of authors (De Felice, 1958, Bouisset *et al.*, 1959; Bouisset and Monod, 1961; Wisner and Rebiffe, 1963; Wisner and Monod, 1963). These measurements vary in particular with age, nationality and socio-professional status (Bouisset, 1981). This is why it is recommended that they should not be used without verification that they correspond to the sample of operators working at the station.

The maximal prehension space defined by the envelope of extreme positions occupied by the hand during a movement of rotation about the shoulder is particularly useful in determining both the placing of the commands to be given by the position of the hand on the master arm handle, for example, or those involving the action of the fingers on, for example, electric switches controlling movement. The maximal space of prehension varies according to the type of prehension, the orientation of the hand and posture.

Optimal zones can then be defined, corresponding to the extreme positions occupied by the hand while each joint of the upper limbs explores the limits of the ease angles (Rebiffe *et al.*, 1966). Bouisset and Rebiffe (1978) and Bouisset (1981) applied numbered precision to workstations involving only light work and with no particular visual constraints. Davis and Strubbs (1977) defined the zones reached in the manipulation of loads weighing more than 10 kg, taking into account the intra-abdominal pressure of 90 mmHg, which constitutes the threshold for the risk of injury. Their study indicates that to exert a thrust in the horizontal plane using the upper limbs without exceeding this risk threshold, the distance limit between the shoulder joint and the placing of the hand is reduced when the force exerted is higher. The maximal thrust that an operator can exert with the arm fully extended in the same plane as this is 18 daN, but if the arm is fully flexed at the elbow such that the distance from shoulder to hand is only one-eighth of the maximal distance, the possible thrust is 30 daN.

Protection of the lumbar-sacral curve is very important and deserves particular attention. Keegan (1953) established the different aspects of the lumbar curve as a function of posture and its alteration.

The workstation must be designed so as to avoid postures which require considerable muscular effort to maintain them. The action of standing up from a sitting position causes an increase in cardiac frequency of about 20 per cent (Turner, 1927 see Monod, 1981) which corresponds to the additional effort required of the postural muscles and modifications in the circulatory system, requiring additional output from the heart (Monod, 1981). Any posture involving raising of either one or both of the arms above the level of the heart is particularly costly and should be avoided. Generally, it is important to avoid a rigid or asymmetrical posture or at least to avoid having to maintain such positions for long periods of time.

Finally, the design of the control handle etc must correspond to its use. The shape of the handle must be adapted to the hand to allow the best gripping conditions. The various ways that the hand can grip and hold an object have been classified by a number of authors, including Taylor (1954), and each type of grip corresponds to a gripping force.

If two very different types of grip are considered: the *clamping grip* (in which all five fingers close around the object, with the thumb covering the fingers to lock the grip) and the *pincing grip* (in which the distal phalanges are strongly flexed so as to place the tips of the fingers and thumb in opposition), the maximum gripping force varies from 29.7 to 46.4 daN in the first case and from 5.8 to 12.9 daN in the second case. But the force also depends on the position of the handle and the angle between the hand and forearm (Bouisset, 1981): the force is maximal for dorsal extension of the wrist of about 40°; this position is often adopted spontaneously for exertion of force. Handle design must consider both the demands in terms of force and precision of the task, as well as other parameters.

Bilaterally servocontrolled teleoperation leads the way for better definition of the force demanded by the operator and that supplied by the slave arm. Current work on computer-aided (CAT) bilateral servo manipulators will allow easy programming of master arm without the constraints imposed in mechanical manipulators, in which the master arm must have the same geometry as the slave.

3.2.6 Analysis of work in real situations and evaluation of the work load

Most studies on the interaction between operators and manipulators in the last two decades have been carried out in laboratories, and make little reference to the evaluation of the work load in real work situations.

Studies of aircraft pilots using simulation devices have not been modified to allow the operator telesymbiotic feedback of all the kinaesthetic sensations experienced by the pilot of a manned teleoperation vehicle. Synthesis between these two fields could provide the necessary expertise to allow evaluation and improvement of teleoperation systems.

The performance of the man-teleoperator system can only be properly evaluated in a real work situation. The problems facing the ergonomist are complex and far less controllable than those encountered in the laboratory. Any external intervention interferes with normal working conditions and the results of any evaluation. The ergonomist must devise methods of analysis which disturb the behaviour of the system as little as possible, and the choice of methods must also take into account the non-scientific considerations since the operators also have their opinions on the methods used.

The main factors in the evaluation of the operator work load are shown in Figure 3.2.6.(1), which provides an operational definition.

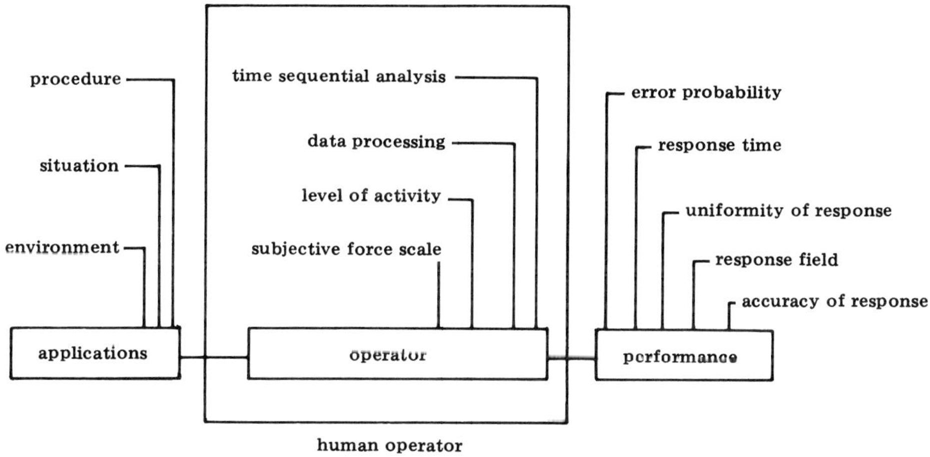

Figure 3.2.6.(1) *Analysis of the operator's work load (from Johannsen, 1979)*

Any evaluation must take into account the three aspects of the man-manipulator system, functionally linked by the role of the operator: the demands of the task, the ability of the operator and the performance (Johannsen, 1979). This implies a comparison of measurements with respect to these aspects, and Table 3.2.1 presents some of the methods available.

Operator demands can be classified into three groups: those that relate to the environment, the workstation design and the work processes and organization. The last group is more important in relatively undefined maintenance work or intervention involving a limited degree of safety.

The demands associated with the environment are those of the physical surroundings. The operator's work is only effective for values of these conditions within a limited range. Excess noise, dim lighting

		remote manipulation stations	types of measurement
		source of demands	
task demands	environment	noise, light, heat	physical measurement
	design	arrangement of station characteristics of equipment, levers, key-boards etc	diagrams of man-machine connections environment sequence of use
	procedures	duration of task, pauses organization of work	systematic observation recording of activities in real time
		types of action	
operator's abilities	physical load	posture, dynamic movement, displacements	analysis of cardiac frequency systematic observation of postures, physical and gestural activity
	mental load	assimilating information	measurement of eye strain
		level of caution	observation of eye movement
		subjective view of abilities	questionnaire
		performance characteristics	
performance	errors	dropping objects grasping too effective/not effective enough	observation
	response time	duration of elements of task	measurement of duration
	precision	precision of positioning trajectory error	frequency of errors

Table 3.2.1 *A comparison of measurements with respect to the task demands, operator's ability and performance*

and considerable vibrations will cause a deterioration of operator performance in the man-machine system, and may have undesirable consequences on health. In man-teleoperator systems, these effects are accentuated and may become dangerous during work carried out a short distance away from a space capsule or submarine, or during remote work in a hostile environment using a protected teleoperation system (eg radioactive environment, handling explosives).

The workstation and equipment design imposes constraints on the operator. The position of the different monitors (connected to television cameras or those supplying other data), the quality of the glass windows and the mechanical properties of the remote manipulators are examples of factors that contribute to the work load. There are also factors associated with the work process and the organization, such as the time required for task execution, the variety of different types of task or the rigidity of the sequences of actions in the execution of certain manoeuvres.

The operator's abilities linked to the demands of the work, described above, are estimated on the basis of measurements of the physical and mental abilities. Physical capacity can be estimated objectively by measuring such factors as cardiac frequency and oxygen consumption (Monod, 1981). The mental effort depends on the operator's condition, and the amount of data to be assimilated. These factors are often difficult to assess, since they can depend on a series of stable factors such as the level of training, and variables such as attention, motivation etc (Johannsen, 1979). A subjective test is frequently the only practical means of evaluating the mental work load.

The performance of the operator in the system is also a contributory factor. The results of the task are generally considered as performance measurements, but the performance should not be the only measurement taken for the work load. The operator can often compensate for increases in the work load by a temporary increase in effort, maintaining performance at the same level. When comparing remote manipulators, performance is not sufficient, and the ability of the operators must also be considered.

Performance measurements are decided by the type of task. The most common performance measurement relates to the speed of execution of the task (eg duration, number of objects manipulated in a given time) or precision (eg number of positioning errors, collisions between parts). The measurements used in the ergonomic evaluation of a workstation depend on the situation studies and the aims of the study. These often result in performance measurements with respect to the different tasks or situations (see Section 3.1.3.4).

3.2.7 Performance evaluation of remote manipulator systems under different experimental conditions

Any study aiming to evaluate a system's performance requires to be tested in a laboraory, and an analysis of work in a real situation can provide specific points relevant to the experimental work.

An in depth knowledge and strict control of the variation factors are important, and, as has already been stated, remote manipulator performance cannot be considered independent of the man-manipulator-environment system.

Various factors, such as age or motivation, may influence operator performance. Environmental factors such as temperature can also affect the level of attentiveness, and factors of this type can modify the results of an experiment in which insufficient controls have been imposed.

The factors which may have an effect on the performance must be carefully defined, and used as the basis of the test task. These experimental factors can be varied systematically to allow observation of the consequences on performance (see Section 3.1.3.4).

Finally, performance must be defined using simple measurements with sensitivity to the experimental factor allowing any phenomena to be revealed. Many authors have attempted to characterize manipulators by measuring force or velocity etc. Vertut and his group based their classification on the dexterity achieved in a variety of test tasks, with measurements sensitive to these factors.

Vertut (1973I, II, III, IV) defined simple tasks for grasping, displacing and inserting objects, and simple assembly processes, or more complex versions. Performance standards were shown by the ratio between the durations for performance of the tasks using remote manipulators and the bare handed times. In this way, he was able to classify different types of manipulator as being capable (or not) of increasing the operator's dexterity. In particular, he established the sensitivity of the dexterity to the quality of feedback to the operator under good viewing conditions.

McGovern (1974) and Hill and Salisbury (1977) attempted to verify Fitts law on the task Id with respect to the task execution time. This involved a test task of transporting an object and inserting it into a shaft, with Id being varied by the degree of tolerance of insertion. The results are partially compatible, but do not provide a clear response to the verification of the law in question. But since the number of subjects was on average 2 and the number of tries ranged from 2 to 35, it would seem that application of the method was not sufficiently rigorous.

Although McGovern (1974) and Hill and Salisbury (1977) found significant differences between the durations of execution of a simple task using two different remote manipulators, the very different design of these systems, as Book and Hannema (1980) note, makes it difficult to draw any real conclusions. In particular, the effects of the dynamic characteristics on performance cannot be guessed at, and such results are of limited use.

This is not the case with Vertut's (1973I, II, III, IV) results which are of general use because they are based on wider and better defined principles. Various other studies comparing modes of control of remote manipulators (Nevins, 1972; Vertut, 1976) have defined:

1. *integrated positional control* using a master slave with or without force feedback;

2. *integrated velocity control* using a single handle (syntaxer) with or without force feedback;
3. *analytical unilateral velocity control* using push-buttons and handles.

A task is executed more slowly when the remote manipulator is subject to velocity control than when control is positional. In this case, force feedback improves the velocity of execution still further. Vertut's results shown in Figure 3.1.3.(3) are in agreement with those of Muller (1973) who indicates that performance is four times faster when a master-slave remote manipulator without force feedback is used than when it is subject to velocity control. Wilt *et al.* (1977) distinguish between small and large remote manipulators. In contrast to small manipulators, large manipulators allow the prehension space and the force of the operator to be amplified, but this is dependent on the view of the work space and the machine's response. These remote manipulators with spatial gain and high energy levels are less sensitive to the control mode: the difference between the performances achieved with the two modes of control is less significant.

Out of the many authors who have attached importance to the influence of different types of signal feedback, feedback transmission delays and feedback transmission modes on performance, Ferrell (1965) holds that the deferred signal feedback has a negative effect on the human performance when movement is carried out in the closed-loop mode. In a later experiment with a force feedback system which lacked visual feedback, Ferrell (1966) observed that when the transmission delay for the force feedback is long (10 s), the manipulator becomes unstable. This can be eliminated if the feedback is not presented to the operator via the hand moving the control handle. Micaelli (1982) during work at Saclay succeeded in stabilizing such a system using the high-level control mentioned in Section 1.5.4.

Bejczy (1975) compensated for the absence of a local view of the work area by using proximity sensors to locate the gripper, and these data were transmitted to the operator in the form of four different sounds. This system reduced the number of errors by half.

Espiau and André (1982, 1983) provided assistance using computers and proximity sensors to help with prehension by the intermediary of force feedback. The evaluation of this system was planned for 1983.

3.2.8 Comparison of the different models of mechanical master-slave manipulators in experimental situations

The study reported here was carried out at the request of the CEA. It consisted of a comparison between the performance of two light duty manipulators (5 kg) and two heavy duty models (20 kg). All four were mechanical master-slave manipulators.

It was decided to compare the performance that the operator could achieve with each of the systems and the physical work load suggested for each manipulator.

3.2.8.1 TO DEVELOP AN EXPERT KNOWLEDGE OF THE DIFFERENT REMOTE MANIPULATOR MODELS

This was achieved using plans and mechanical properties provided by the manufacturers to allow the work zones and performance within these zones to be established for the remote manipulators. These data were then compared with those available on the work zones of operators (Bouisset, 1981; Davis and Strubbs, 1977).

There is no strict relationship between work zones established in this way, and this has a direct consequence on the performance that can be expected from the man-manipulator system, and must be considered when arranging the work area. The modification of the ratio of forces between master and slave affects the quality of the data received by the operator and the operator's responses, but it was not possible with these models.

3.2.8.2 STUDY OF THE SITE

This usually takes the form of a systematic study of the workstation both in production and maintenance, in particular with operators. The study also includes a bibliography. The many variations of workstation prevented an accurate analysis of the site, and the team could not perform this pre-analysis with sufficient thoroughness. It was, however, possible to determine the experimental characteristics necessary for the laboratory: the tasks, the variables measuring performance and work load, the choice of subjects and the experimental device.

3.2.8.3 EXPERIMENTAL TESTING

The tasks

Two tasks were chosen so as to explore all the possibilities of remote manipulators and reproduce the types of task executed in production. The first was the transport and positioning of objects (cube and cylinder) on targets situated at different heights in the work space [see Figure 3.2.8.(1)]. The second task involved precision: the introduction into a box of four different shaped parts into apertures of corresponding shape, unscrewing the cover of the box and removing the shaped parts one by one [see Figure 3.2.8.(2)].

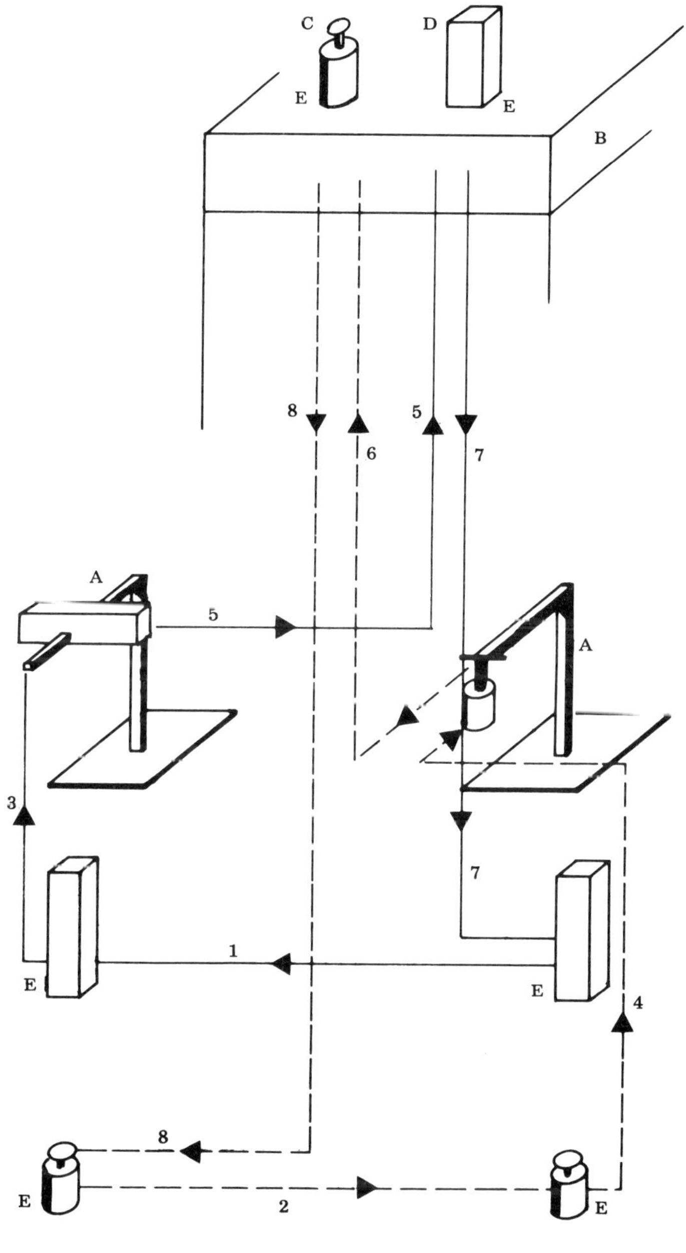

Figure 3.2.8.(1) *The steps in task 1 using a 20 kg manipulator:* **A**, *winches;* **B**, *bench;* **C**, *cylinder;* **D**, *parallelipiped;* **E**, *targets on work surface and bench. The trajectories are numbered in order. At each displacement the operator adopts a posture then changes to another*

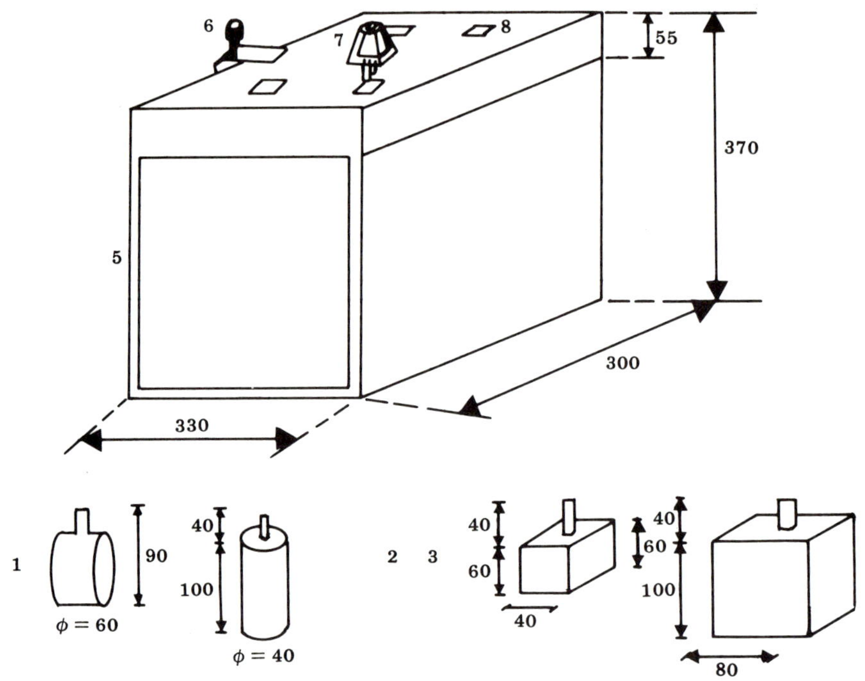

Figure 3.2.8.(2) *Task 2: box and inserted parts* **1 to 4**, *inserted parts weighing between 250 and 600 g, dimensions in millimetres;* **5**, *box with transparent sides;* **6**, *screw to open box (three turns);* **7**, *handle for removing cover;* **8**, *openings corresponding to loose parts for insertion*

The variables

The variables intended to measure the performance and effects of the work load were:

1. the length (in s) of the execution time of the tasks;
2. the length (in s) of the execution time of each phase of the tasks;
3. the identification and analysis during execution of the following errors:
   - letting the object drop;
   - fumbling, hesitation: the operator fails either to carry out the correct displacement or to deposit the object;
   - errors in positioning the object;
   - other;
4. the recovery pulse, that is the cardiac frequency in beats per minute at rest, 30 s, 1 min and 1 min 30 s after execution of T;

5. the systematic identification of postures during execution of the tasks according to two techniques:
   - the method described by Owas, cited in Karhu (1977);
   - the recording on videocassettes of the subjects running the tasks, and the analysis of the following postures, classified beforehand:
     (1) arm above heart level;
     (2) arm below heart level;
     (3) flexion of lower limbs;
     (4) straightening of lower limbs;
     (5) flexion of trunk;
     (6) extension of trunk;
     (7) torsion of trunk;
     (8) lateral flexion of trunk;
     (9) extension of neck;
     (10) flexion of neck;
     (11) rotation of neck;
     (12) support forward of operator on manipulator (posture adopted to compensate for moment of slave arm);
     (13) support of remote manipulator on shoulder
     (14) support of remote manipulator on hip;
     (15) support of remote manipulator on thigh.

The last three positions compensate the balancing faults of the remote manipulator or the momentum of the handled load when the slave arm is oblique [see Figure 3.2.8.(3)].

The subjects

The differences between subjects are a source of important variation, for this reason it is important to form groups of subjects which are as similar as possible, in particular with respect to the degree of training in each manipulator, and it is preferable to choose them from production operators.

The experimental system

It must reproduce the work situation, which is why a porthole was placed between the operator and the work bench. The manipulators were all placed in the same position with respect to the porthole so that the conditions of work would be as similar as possible.

Protocol

Four manipulators and two tasks imply eight experimental series. Eight subjects performed them during two sessions in a different order so as to reduce the effects of training and fatigue.

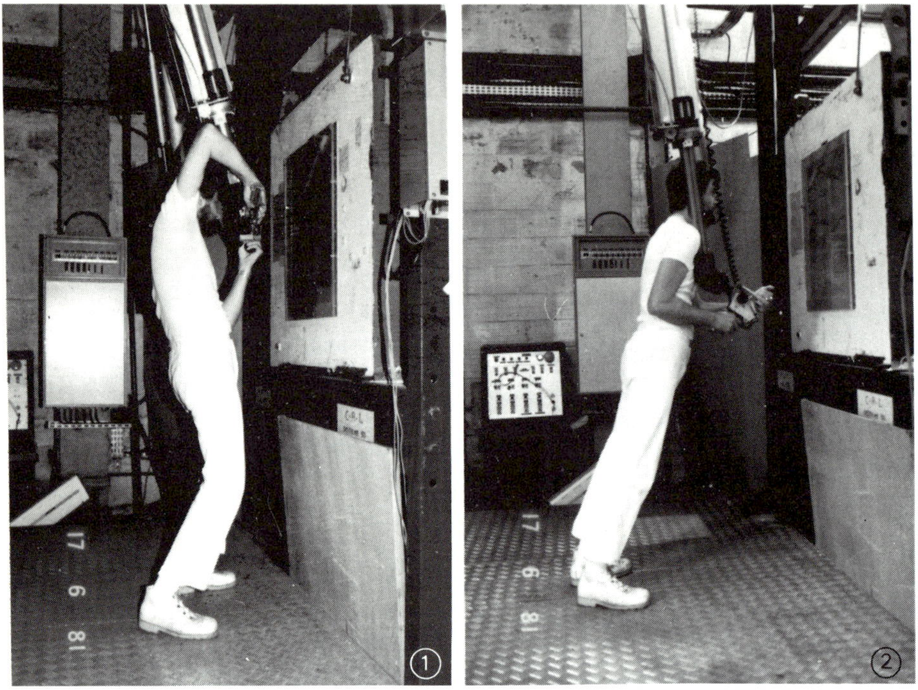

Figure 3.2.8.(3) *Characteristic posture during tests:* **1**, *posture avoided by Z indexing motions;* **2**, *posture of support on shoulder necessitated here by mass to be moved to back of chamber, slave arm oblique and extended in Z (this may be added to a fault in counterbalancing in other models);* **3**, *posture avoided by 5.5 indexing motions of X and Z;* **4**, *posture allowing indicator lamps to be seen showing the possible indexing motions. This shows one of the reasons for the preceding postures linked to insufficient use of indexing motions: an audible signal distinguishing Z, Y and X could solve this problem. Another manipulator with individual switches for indexing motions leads to handle difficulties;* **5**, *another avoidable position, the operator is too far from the window. Y indexing motions and proper dimensional adaptation of the manipulator could avoid this;* **6**, *inevitable position because X indexing motions are not available with this model. This ergonomic study shows the possibilities for improvement which were unknown to manufacturers*

Results and conclusions

A questionnaire was given to each subject, containing questions on the feelings of fatigue, painful areas etc. This demonstrated that the operator presents symptoms which are not specific to the task on any particular manipulator, but more generally associated with remote manipulation as a whole. The developments taking place in control handles should improve conditions regarding this point but real improvements will be brought about by the use of bilateral servo

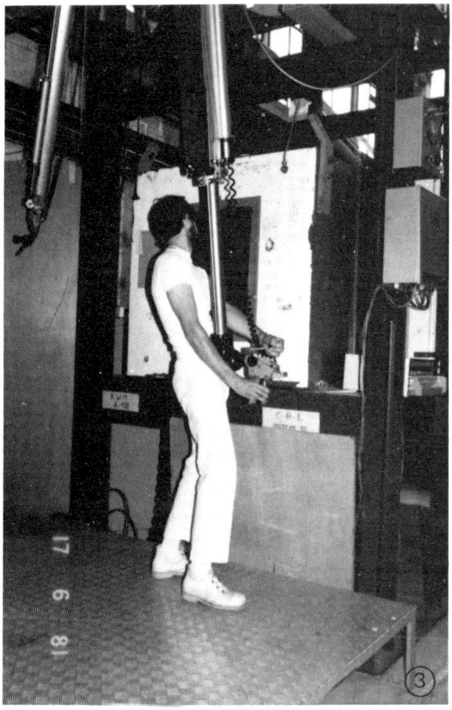
3

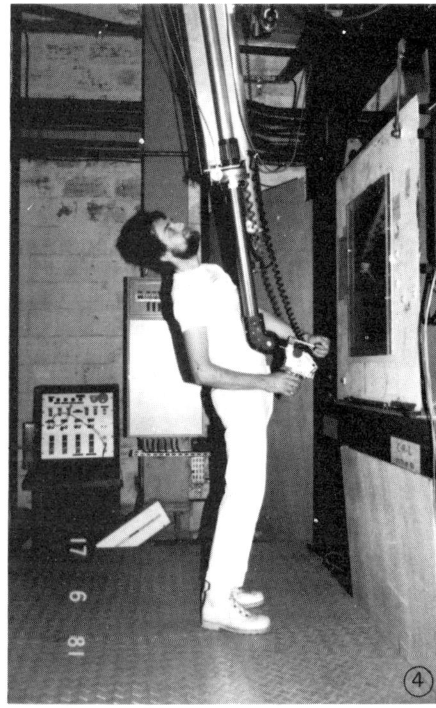
4

5

6

manipulators and the assisted control mentioned above. In the meantime, the best solution is probably to rest for periods during work. It should be noted, however, that the tasks chosen represent an intensive load compared with that usually encountered in real work. It seems likely that a more conventional experimental method would provide less subjective results.

3.2.8.4 COMPARISON OF TWO MANIPULATORS (A AND B) CARRYING A MAXIMUM LOAD WEIGHING 7 KG

The results show great coherence. In performance, any tasks take longer with remote manipulator A, and the number of errors is also larger when the the subject uses A. Among the errors classified, those in positioning the objects were most common. In the work load, the increase in cardiac frequency is greater after execution of the task with A [see Figure 3.2.8.(4)]. The difference is small (about 5 beats per minute) but systematic. Note also that this is more marked after a

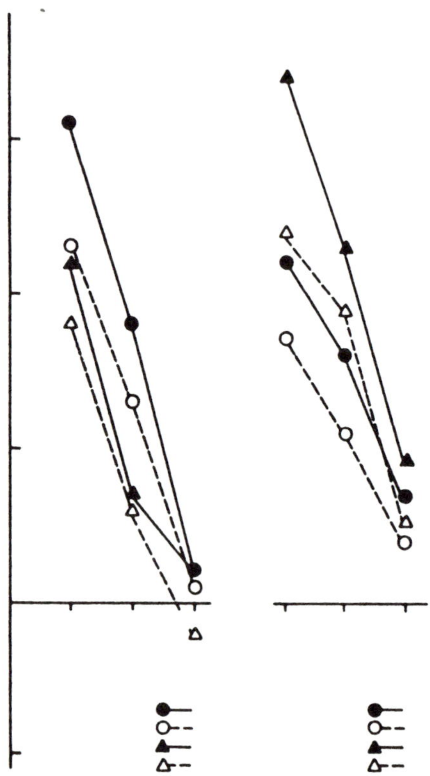

Figure 3.2.8.(4) *Comparison of two manipulators (A and B) carrying a maximum load weighing 7 kg*

transporting task than after a precisional task (15 as opposed to 10 beats per minute).

Postural analysis shows that work with A is more awkward than with B, and that while lowering the object from a shelf to the work bench, the subject adopted a torsional position of the trunk in A, and one of lateral flexion of the trunk while using B. The posture required for unscrewing with A was far more uncomfortable than that required for B: elevation of both arms above the heart, torsion of trunk, extension and rotation of neck.

These results clearly show that remote manipulator B is superior to remote manipulator A, both in the precisional and velocity performance and the work load on the operator, in relation to the cardiac and postural cost. It is not yet known whether automation of the Y indexing motions would alter these conclusions.

3.2.8.5 COMPARISON OF TWO MANIPULATORS (C AND D) CARRYING A MAXIMUM LOAD WEIGHING 20 KG

The complexity of the results demands precise analysis and provides somewhat inconclusive results. In terms of performance, statistical analysis does not differentiate between the two systems, and the number of errors is the same for both. But there are more cases of objects being dropped by C, whereas grasping errors are more frequent in D. This can be explained by the fact that D is equipped with electric gripper closing, and once the object is grasped it is held firmly, but the actual grasping is more difficult. The increase in cardiac frequency is greater after the task when carried out by system D [see Figure 3.2.8.(4)]. The difference is, once again, small but systematic. The analysis of the postures adopted by the subject indicates that the use of D involves a higher postural cost, and while placing and removing an object from the shelf, the subject flexes the lower limbs, flexes the trunk laterally and leans forward on the manipulator, more frequently and for longer than with C, but with C this phase involves more frequent neck rotations. For precisional work, the use of D leads to either or both arms being raised above the level of the heart, whereas the same work with C leads to extension of the trunk and neck, which is a less costly posture than the other.

These results on the work load are interpreted as follows; remote manipulator C is counterbalanced in all three dimensions. There is no Y counterbalancing of system D, and this causes an increase in the moment of the slave arm which the operator must compensate with the exertion of considerable forces. This is particularly true when the slave arm must be placed in the extreme zones of the field of work.

Finally the high frequency of rotation of the neck during work with system C is explained by the eccentricity of the system of control lights

task \ remote manipulator	total duration of task: manipulator C	total duration of task: manipulator D	statistical test: F of Snedécor	total number of errors: C			total number of errors: D			statistical test: t student
transport	17 min 27 s	15 min 2 s	1.48 for 1 and 7 dl; nm significant	50	10	26	54	19	18	0.53 for 7 dl; non-significant
precision	16 min 34 s	14 min 28 s		64		↑	65		↑	0.06 for 7 dl; non-significant
					object drop				object prehension error	

task \ remote manipulator	total duration of task: manipulator A	total duration of task: manipulator B	statistical test: F of Snedécor	total number of errors: C		total number of errors: D		statistical test: t student
transport	8 min 11 s	4 min 15 s	17.67 for 1 and 7 dl; $p < 0.1$	74	36	24	27	5.25 for 7 dl; $p < 0.01$
precision	12 min 29 s	6 min 22 s		56		33		3.05 for 7 dl; $p < 0.05$
					object positioning error			

Table 3.2.2 *Statistical analyses for remote manipulators A, B, C and D*

relative to the operator's field of vision. We suggested that this system be moved [see 4 in Figure 3.2.8.(3)].

In view of these results, it is not possible to draw the same type of conclusions as for the case of A and B, but it can be stated that the work load is less with system C because of the technological differences between it and system D.

Although these results required work to be carried out in the extreme zones of the remote manipulators, the frequency of this type of task in real use justifies the study and the conclusions we have drawn from it. Table 3.2.2 shows the statistical analyses for A, B, C and D.

3.2.9 Ergonomic consequences of the technological advances in teleoperation

The role of the operator in the man-manipulator system has not really changed over the last 20 years despite the developments in the manipulator itself. The progress made in the field of technological constituents since the appearance of the first remote manipulator have only been partly integrated into the design of more recent models. A recent technological evaluation has allowed various areas to be classified in the application of new technological constituents to the design of remote manipulators (Diaz and Boykin, 1979). The work currently being carried out in the ARA project is contributing to the completion of this approach, shown in Table 3.2.3. The main modifications for the operator in the man-manipulator system will probably be based on:

1. technical improvements in the standard systems: articulations of the arm, actuators etc;
2. generalized master-slave control, freeing the master structure from the slave structure, using computers;
3. development of manual control towards computer assistance via extremity sensors on the slave unit;
4. development of manual control towards shared control by man and computer;
5. introduction of improvements in the arrangement of the workstation, particularly with respect to the presentation of data processed by the computer, informing the operator in a synthetic way of the state of the system, and the execution of the task;
6. automatic computer intervention for security reasons preventing the operator from making errors.

Computing allows considerable possibilities. Cooperation between the human operator and the computer promises an improvement in performance, and speed and precision (Starr, 1981). This mode of control, which is studied in several laboratories in the USA, is known as

technical progress	man-machine interface	consequences for operator
actuators	error reduction in manipulator characteristics	improved transparency
structure and articulations	improvement in use of DOF	
presentation data		
— on work area	improvement in video systems, use of sonar or lasers	increased visual feedback in shared or manual control
— on gripper sensors	additional data on proximity, position and force applied to gripper	kinaesthetic data in shared and/or manual control
control systems		
computer transmission of commands from master to slave	possibility of eliminating noise of human movements	improved transparency
control modes: position, velocity	mode adapted to task: fine transport of manipulation	increased speed of execution
control shared with computer	certain elements of task are programmed and executed independently of operator	reduction of repetitive movements but increase in surveillance work
	partial automation in parallel with operator	increase in speed of execution for precise tasks
automation with security	under control of operator or not	
I/O between operator and teleoperator	improvement in handles and development of syntaxers for force with DOF	mode of control can change to force
	polyvalent miniarm	promising solution
	alphanumeric keyboard	mental load may be increased when task modified
	systems with voice control	can sensorial modalities be used simultaneously
arrangement of work-station	development of work for operator and equipment	increase of constraints of reduction in space; possible specific adaptation to VDU

Table 3.2.3 *Technological evaluation of remote manipulator systems*

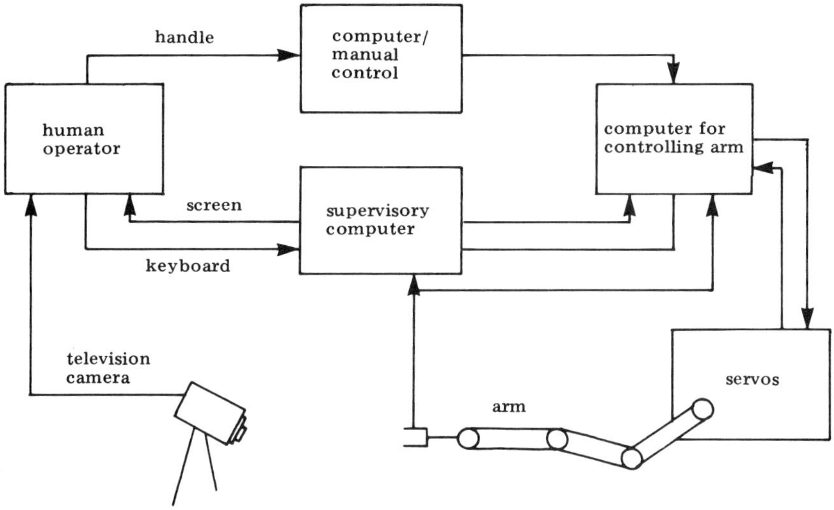

Figure 3.2.9.(1) *Supervisory control*

supervisory control (Starr, 1978; Sheridan, 1976; Bejczy and Zawalki, 1979). Figure 3.2.9.(1) shows this type of control. Operator participation in this mode of control is on four levels: planning, training, surveillance and intervention (Sheridan, 1976). During the planning stage, the operator defines the movements of the slave arm, away from the control station without time constraints. The planes are then introduced into the computer. This constitutes the training phase. Surveillance consists of watching the various systems for data presentation with respect to the movements of the slave arm, controlled by computer. The operator can also control directly, and manually, the remote manipulator.

Preliminary experiments carried out by Starr (1978) have led to improvements being made in assembly tasks with reduction in the transmission delays. This method is still being developed, however, and the role of the operator is still not fully defined. This approach will probably be used for teleoperation in space with transmission delays of over a few seconds. The approach to these ideas in the ARA project has led to a precise analysis of the different modes of control shown in Figure 2.1.4.(4), in which the concept of supervisory control is contained within the large framework of advanced teleoperation. The differences with Figure 3.2.9.(1) are clear, and without going into detail on 2.1.6, Figure 3.2.9.(2) shows the breakdown of the different functions of the operator and computer (which is not physically made up of three calculation units). The relationship is clear, in particular

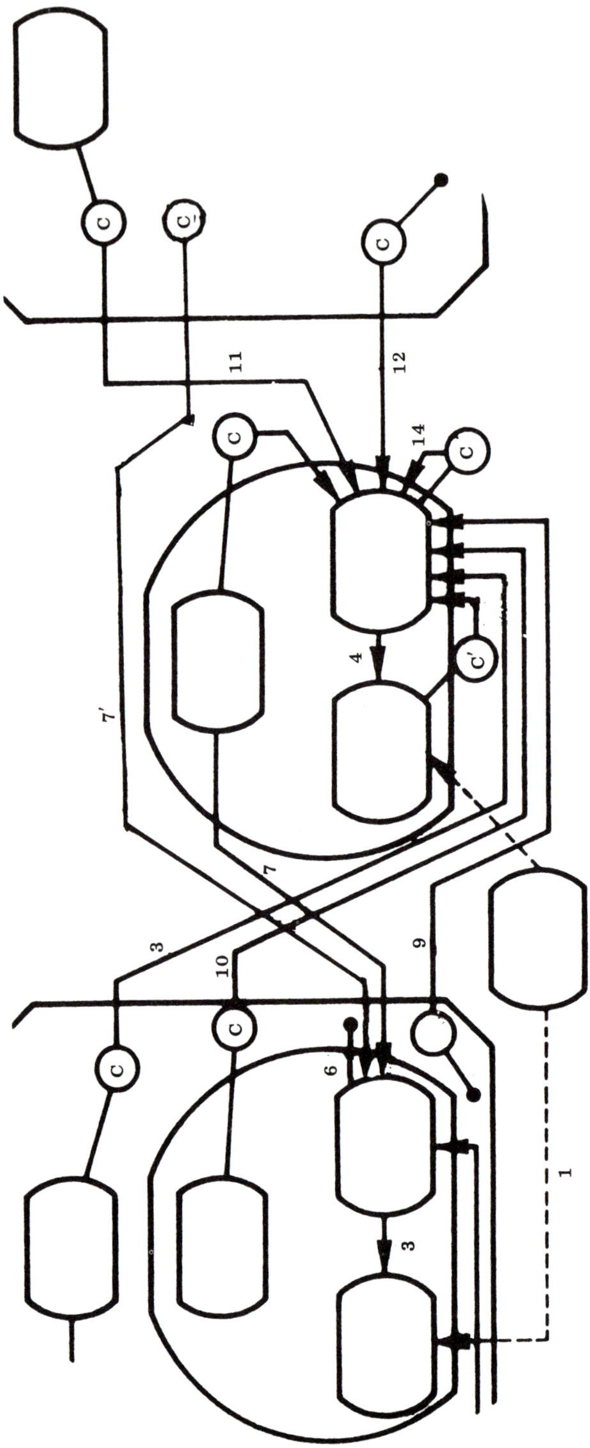

Figure 3.2.9.(2) *Breakdown of the different functions of the operator and computer*

in the form of the sensing computer-sensing man dialogue, and the contribution of data processing by the computer is clear. The attribution of various tasks to the man or the computer, either on command by the operator or request of the computer, will open up entirely new developments and the definition of criteria allowing automatic transfer to computer control from operator control without risk of conflict between the two leads onto advanced teleoperation with artificial intelligence.

There are of course new ergonomic problems associated with these developments, but the ARA project intends to introduce ergonomics during the development of these new modes, to avoid any inefficiency in deployment. This as much concerns the physical links between operator and computer as the development of appropriate control languages.

The planning phase is considered in the ARA project with a view to the importance of work which is difficult to model in advance, such as emergency intervention or maintenance. It is hoped that the resources of CAD will be used in this, or for partial realization of the task. In this, advanced teleoperation is moving the same way as robotics in general, towards greater autonomy of systems.

One result of these developments will be a reduction in the physical work required of the operator, but an increase in intellectual effort at the planning phase. It is, however desirable to reduce the amount of mental work required of the operator too, without there being any prolonging of the planning or execution stages, and the ultimate, but ambitious, aim is to reduce both these times and the work load also.

New handle design must take into account the characteristics of normal human movement in execution of the manipulation tasks, and help suppress the parasite torques which are currently unavoidable.

Compound modes of bilateral positional and velocity control limiting the need for movement in the hand of the operator will improve the gestural language of robot and teleoperator control while other control possibilities are being studied. Among these other ideas is that of word recognition to transform spoken commands into computer commands, and this is also the case in the more specific field of control of prostheses and ortheses using signals from the muscles of the face (Brenguier and Bouisset, 1981) or movements of the head (Quetin, 1978) for controlling a slave arm.

These modes of control are less fast than traditional manual control and will probably remain exclusively for the control of large transporters or vehicles by able-bodied operators and teleoperator systems for handicapped people.

All these advanced techniques must be evaluated as much by ergonomic criteria as by engineering criteria. The effects of each must be considered with respect to the performance of the man-machine system and in relation to their consequence on the human operator's work load.

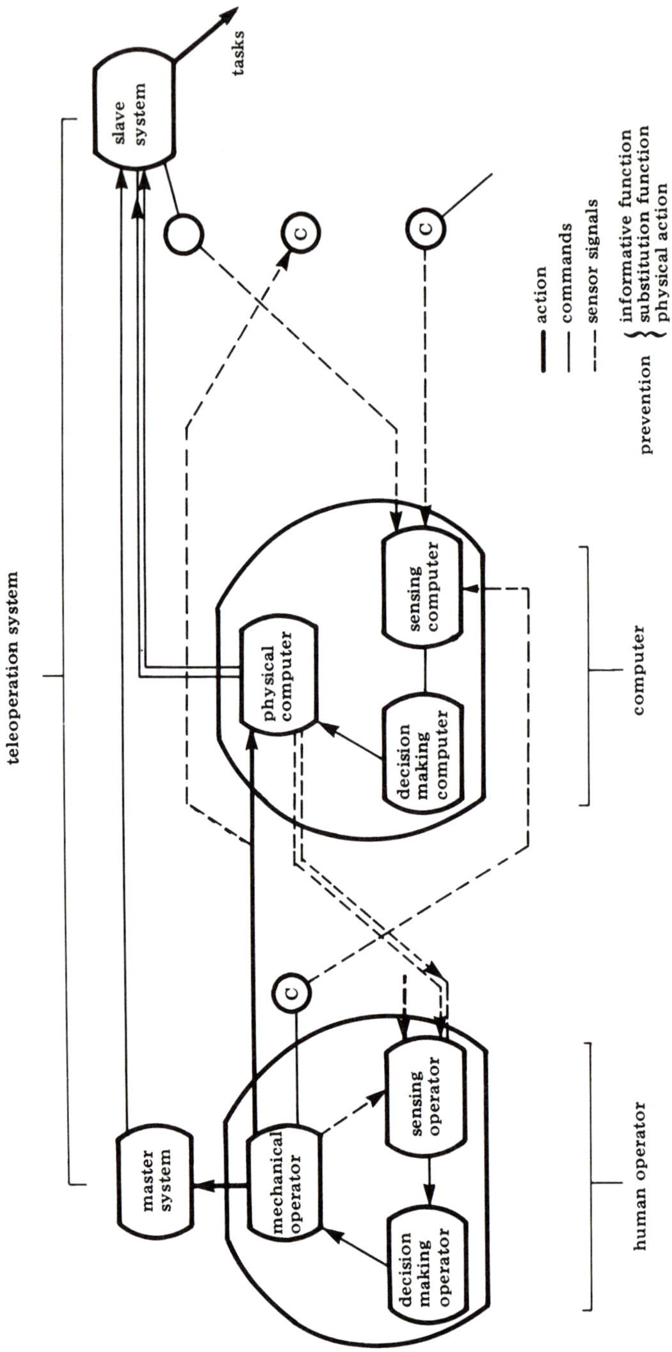

Figure 3.2.9.(3) *The function of the computer in the teleoperation system*

Part 4
Applications of teleoperation

Introduction

Most of the applications considered in Part 4 have been mentioned in earlier chapters, but is seems appropriate, at this point, to make an appraisal of the current progress in engineering in this field and to speculate on the prospects in each category of application. The intention of Part 4 is to generate a conclusion to the work as a whole, by briefly reconsidering the material from earlier parts of Volumes 3A and 3B, organized in terms of application.

The approach will emphasize the importance of a common scientific and technological basis to the many branches of this technology. This has been absent from teleoperation in the past, and, with the possible exception of nuclear teleoperation, is still lacking today, but will be indispensable in the growing field of advanced teleoperation. Links between advanced teleoperation and advanced robotics will be stressed, and the importance of these links for the future of both fields forms a major part of the conclusion.

Chapter 4.1

Nuclear applications

4.1.1 Research and pilot facilities

This is the field in which mechanical master-slave manipulators and motorized remote manipulators have been most widely used. Mechanical master-slave manipulators are able to perform, on a laboratory bench behind a heavy protective shielding, standard tests involved in research including delicate manual operations. Remote manipulators are also used for the assembly, removal (or dismantling) and maintenance (or replacement) of all the required experimental equipment.

This is also the case with research work concerning non-destructive examination of irradiated or spent fuel, metallurgical research on radioactive alloys (and particularly the behaviour of metallic or ceramic fuel), and radiochemical work on radionucleotides used in medicine and agriculture, or the reprocessing of irradiated fuels (particularly in pilot reprocessing facilities) (Faugeras *et al.*, 1963; Lefort *et al.*, 1964).

4.1.1.1 HOT CELL ROWS IN NUCLEAR RESEARCH

Photographs of nuclear research hot laboratories generally show mechanical master-slave manipulators lined up along a shielding wall which forms the common work face for a row of cells, each separated from the other by a protective shielding. This shielding is usually made of lead or sometimes cast iron for handling small or medium amounts of radioactivity, and concrete for handling higher activities. Barriers of 1 m thickness ordinary or high density concrete (baryte or magnetite base) are common. This arrangement of manipulators and cells is often in a U-shape, thus enclosing a rear area common to all cells which may face onto a *decontamination* cell, for cleaning or contamination reduction by physical or chemical methods.

The front area, occupied by the human operators, does not need any further protection, and the radiation dose rate is such that work can be carried out there for unlimited periods of time. On the other hand, the rear area, depending on the type of work being carried out, may require protective clothing as slight atmospheric *contamination** may be caused by the presence of a radioactive substance in part of the environment (PMDS, 1982), and the radiation dose rate may be such that only a limited period of time can be spent in the area. The same applies to the front area during removal of a manipulator, if this is carried out from the front.

Associated with each pair of manipulators is a lead glass window, and in some cases a single cell has facing manipulators and windows. In this situation there would be no rear area.

For this type of configuration, a large hot cell is used with a central area for storage which is accessible only to lifting devices and/or motorized remote manipulators, while a zone of about 2 m along the two main walls is covered by dextrous mechanical master-slave manipulators.

Among the tasks indirectly linked to experimental work, those entrusted to the remote manipulators involve transfer operations without any break in the protection of waste removal or for the supply of materials to the facility. It is clear, therefore, that remote manipulation and transfer are two activities that are strongly linked in the design of experimental nuclear facilities, and this is true for all areas of application [see Figure 1.6.1.(2)].

This has led to an approach in which the aim is to optimize the use of teleoperation and remote and protected transfer. This is currently being developed in France by Jean Vertut, and is known as the *teletrans concept* (see Sections 4.1.2.1 and 4.1.2.2) intended for industrial remote maintenance (*telemaintenance*) (Vertut *et al.*, 1984V).

For work involving chemicals, it is usually possible to set up and dismantle (or modify) the experimental equipment using remote manipulators, although the obvious guidelines relating to the flexibility of facilities are not always observed, and this can cause considerable problems when repairs or changes to experiments become necessary. It will be systematically shown in Part 4 that teleoperation and flexibility cannot be dissociated.

The rows of hot cells themselves sometimes constitute the only risk of radiation. This led to the introduction of *sealed sources*, which prevents any contamination spreading.

This is also the case in radiation facilities, such as those used industrially for the sterilization or treatment of a variety of materials. In most cases, however, the materials to be handled come in dispersable solid, powder or liquid form, and this can involve high levels of radioactive contamination in the hot cells. In many cases, a secondary containment enclosure is used inside the shielded enclosure itself, thus separating the containment and shielding functions. Containment only

***Radioactive material strongly integrated into non-radioactive solid material or sealed into highly resistant inert casing (PMDS, 1982).**

is known as a glove box, and comprises long sleeved gloves attached to the shielding in such a way as to be absolutely airtight.

These glove boxes allow work to be carried out on sources with non-penetrating radiation, such as plutonium and other alpha particle emitters.

Remote manipulators and associated transfer methods are the key to successful operation of experimental facilities, because they are directly involved in the repair of breakdowns or changes in the experimental process, and are invaluable during periods of inactivity.

As has already been shown in Figure 1.3.0.(2), and in Sections 1.3.4.6 and 1.3.6, the removable slave arm of sealed manipulators can be a large piece of equipment to transfer to the repair or decontamination areas, and often determines the size of the transfer equipment. The optimization of remote manipulators, their windows and transfer equipment is, therefore, the key to a good hot laboratory design, in which nearly all the existing mechanical master-slave manipulators and motorized remote manipulators (some 20,500 in all) are used.

4.1.1.2 PILOT IRRADIATED FUEL REPROCESSING FACILITIES

These pilot facilities are used to allow processes developed in the context of research to be perfected with a view to their industrial application, but on an intermediate scale. They differ from full-scale plants in the size of the equipment and amount of radioactivity in them. They generally involve a more frequent use of power manipulators because large pieces of equipment have to be moved fairly frequently. Such facilities have provided the inspiration for many developments in teleoperation, because they pose larger problems than research lines. It was at one of these plants that the bilateral servo manipulators, developed by Goertz and known as his model E3 (Goertz *et al.*, 1961I), was tested in 1960. Goertz hoped that, as a result of these trials there would be a widespread development of this series of manipulators, but unfortunately this was never realized. The reason for this is that these facilities tended to be designed by engineering teams, which are invariably more conservative than the researchers, and the size of their budgets discouraged them from taking any risks; because of this the designers of these facilities delayed by some 20 years the introduction of tools that were considerably faster than the unilateral power manipulators originally installed. Since the operational output of these facilities cannot be clearly measured, it has not really been possible to compare the efficiency capacity and flexibility of bilateral servo manipulators with the original systems.

The largest of these pilot facilities will already qualify as a small plant, and, in particular, the Oak Ridge team (Feldmann and White, 1980; Feldmann and Hamel, 1984I, II; Advanced Fuel Recycle Program,

1978) has developed an approach based on wide use of bilateral teleoperation called the *Remotex concept* and now *Teletec*. The prospects for these facilities are currently looking more hopeful than ever before, and these will be considered in Section 4.1.2.1.

4.1.1.3 PARTICLE ACCELERATORS

The five pairs of bilateral servo manipulators currently in permanent use for maintenance are divided among the European Centre for Nuclear Research (CERN), the large Fermi accelerator near Chicago and the linear meson accelerator in Los Alamos. These facilities involve the high targets and other delicate apparatus, which require high-standard teleoperation for maintenance and adjustment, although the method used until recently has consisted of waiting for a sufficient decrease in radioactivity* before carrying out direct manual intervention.

Our colleague, Carl Flatau, first became involved in the field of servocontrolled teleoperation as a result of his thorough knowledge of accelerator problems (Flatau, 1965, 1969, 1970, 1973, 1977; Flatau *et al.*, 1972). Whereas users and designers of hot laboratories are accustomed to the use of windows, and prefer it to the introduction of television cameras, accelerator users, for whom the introduction of windows is inconceivable because of the conception of their machines, have found it much easier to adapt to the use of cameras. This has resulted in very flexible teleoperator systems with servo manipulators mounted on large transporters that fit the space in the facility. The open space available at Los Alamos gave rise to the Monitor [see Figure 1.6.5.(1)], and the restricted space in the tunnels at CERN inspired the Mantis, whereas the Fermi laboratory makes use of a real hot cell and a large Cartesian bridge-mounted transporter. Work on computer-aided teleoperation (CAT) is partly based on information provided by the use of these machines, which are still very few in number, but there is, as yet, no application of CAT in current accelerator facilities.

4.1.1.4 EXPERIMENTAL REACTORS

Generally speaking, these are pool-type reactors in which access can be gained to almost all parts. These facilities are most suitable for direct remote manipulation tongs and rods, which may be 7 or 8 m long. These systems have low levels of dexterity, but direct vision is possible because of the transparency of the water. To the best of our knowledge, there has been no application of dextrous remote manipulation in this

*The induced radioactivity decays partly in the first few days after shutdown. The time required for half the activity to be lost is called the *period*. It is short for certain activated elements.

field, despite the efforts of Michel Grenon's group (Grenon, 1968) in the Euratom programme.

With regard to the future, however, there is currently much reflection taking place on the subject of experimental nuclear fusion reactors with magnetic containment (Tokomak). The first machines intended for plasma experimentation did not require any real effort as far as remote manipulation was concerned, but this is not the case for the larger systems that started operating in 1983-1984, including the Tokomak Fusion Test Reactor (TFTR), Princeton, USA and the Joint European Torus (JET), Culham, Great Britain. The design of these machines has been carried out with a view to making use of bilateral servo manipulators, on vehicles, on the ground and under overhead cranes. This work was led at the JET by Tulio Raimondi, whose name appears on the first Italian publications concerning the Mascot manipulator [see Figure 1.1.2.(10)]. The maintenance of nuclear fusion reactors is discussed in Section 4.1.2.2, with assumptions being made on their industrial nature in the same way as for fuel reprocessing facilities.

4.1.2 Operation and maintenance of industrial nuclear facilities

This section once again outlines the difficulties of moving from laboratory applications to industrial applications. Since these problems are particularly evident in teleoperation, the subject is handled here with caution.

4.1.2.1 NUCLEAR FUEL REPROCESSING FACILITIES

As with the pilot plants that precede the full-scale version, these facilities are designed by engineers whose extreme caution with respect to new technology can be explained by the economics that accompanies the construction of advanced industrial facilities. The result of this is that, in common with the underwater petrochemicals industry (which considers itself advanced), the nuclear industry has a tendency to reject innovation, which could be of considerable benefit, on the basis of prudence.

Engineering does not appear to distinguish between the transfer to industrial applications of new techniques which have just been completed, and that of techniques already used for the same function in laboratories or pilot plants. For example, remote manipulators have been used less in the nuclear reprocessing pilot plants than in research laboratories, and even less in actual operating factories. The enormous degree of imagination and practical knowledge which led to the development of research facilities, and is illustrated in the 33 volumes of

Remote System Technology (*HLE-RSTD*, 1952 to 1984), which extensively appears in the Bibliography (32 years of experience from an international community of hundreds of users and designers), seems to have been totally ignored by those responsible for building factories, and those who give priority to equipment problems without considering or making use of the knowledge and constraints associated with the remote disassembling of the process equipment and the accessibility of this equipment to remote manipulators.

The result of this is that the increased requirements of the process equipment cancels out any advantage of using teleoperation techniques based on dexterity and flexibility, and may decrease the availability of the facility.

There are many components developed with great care to last the entire life of the facility, which are inaccessible and sometimes impossible to remove. When faults arise (or during final dismantling, which may become a necessity) this can present considerable problems that were never considered at the design stage. These situations also arise in reactors.

What is more, in a country where the experience gained in reprocessing facilities is among the best, the practical knowledge was used to select apparatus which needed no maintenance, thus allowing savings to be made in the construction (although less than the extra cost for final dismantling), by rejecting expenses needed for accessibility and teleoperation. Inversely, mechanical master-slave manipulators designed for laboratory use tend to be used for the relatively frequent maintenance of the less reliable components, although insufficient funds do not permit research and development of such manipulator systems for this new use. A compromise is made by simply increasing their capacity, while hoping that performance standards will not be greatly affected, although their design has been improved from the point of view of their ability to be disassembled into three parts, and their mechanical sealed transmission. To the best of our knowledge, the only mechanical manipulators to be designed in the light of reasonably well-defined constraints for their use in the maintenance of a reprocessing facilities are the recent MT200 system by La Calhène [see Figures 1.3.0.(2), 1.3.4.(13), 1.3.4.(17), 1.3.4.(18)] and the corresponding British models.

As a result of the excessive maintenance required for the recent remote manipulators, which were misused in the way described above, there has been a tendency to complicate the design of machines so that they can be remotely dismantled into modules, using a simple bridge crane. This is why the MA23M [see Figure 1.7.4.(2)] has not been used industrially yet in France, and why its introduction has been delayed by some ten years, and there is no way of guessing when its possible successor, the MAE200, will be put to proper use. The same process has also happened in connection with the Frigg offshore oilfield, and the culprits this time are the French oil companies.

This illustrates the way in which the progress of a technology leads to rejection of the progress made in service technologies which could be of use to it.

A completely different process has been adopted by the Oak Ridge team who, in 1978, launched the Remotex concept and designed an experimental plant HEF as part of the project, intended for the reprocessing of Fast Breeder Reactor (FBR)* fuel. This started from a completely different point to that of all other processes of the same type, with accessibility to and availability of advanced teleoperators suitable for this environment. The Remotex concept offers bilateral servo manipulation its first opening in industry. It is true that the team which undertook this project did not start with the engineering skills, but it has the back up of 33 years of *Remote System Technology*, and took the precaution of imposing all the constraints of accessibility and dismantlability etc at the industrial level. The concept is shown in Figures 1.6.3.(1), 1.7.2.(2) and 1.7.4.(1).

The facility consists of a single very large shielded enclosure, with only one entrance, used both for the delivery of materials for processing and the removal of treated materials (plutonium oxide, used as a basis for the production of FBR fuel), thus fulfilling, from the outset, an important safety requirement (in the security sense as well as for protection against radiation), by having the minimum penetration to the facility. Once closed, this huge shielded enclosure is also completely contained, with clean equipment only entering through the decontamination chamber, and the entire reprocessing takes place using an internal means of remote manipulation. These teleoperators account for the decision to protect one large volume instead of isolating each piece of process equipment with independent radiological protection. By this means, the amount of concrete needed to build the plant is reduced, and at the same time its strength is improved, because there is no need for windows, and the savings made are passed onto the teleoperation equipment itself. This cannot be justified unless, at the end of these efforts, the availability of the plant is improved. The risk of contaminating some of the process equipment in the event of a failure involving another unit is accepted in this concept, but the teleoperators provide direct access, and intervention can be carried out very rapidly. In this way, by applying the knowledge gained in research hot laboratories, it is possible to avoid undismantlable equipment, and the modular construction of equipment, which is easily repaired or replaced, can be encouraged.

This is a summary of the principle of Remotex remote maintenance, which also provides the potential for development during its use, a means of repair in case of major accidents and, when the plant is shut

*The FBR is a sodium-cooled reactor using fast neutrons *burning* plutonium fuel. The fast neutrons generate more plutonium from non-fissile uranium located around the fuel, called the *blanket*.

down, the same teleoperation methods can be used for final dismantling and decommissioning.

This type of approach involves a tremendous rebalancing in the proportions of teleoperators and process equipment, while teleoperators and transfers account for some 2 to 3 per cent of costs incurred in fully operational factories.

This concept, if it is adopted industrially, will be a highly significant factor in the progress of advanced teleoperation. The current state of the work at Oak Ridge only includes a computer application for digital communication and servocontrol, allowing the system to operate with a reduced number of wires or by wireless link.

The extremely high levels of radiation present in such plants mean that considerable progress has to be made in the radiation hardening of the components, although all these changes are entirely possible and will only require coherent research and development programmes to bring them about. The extensive use of electronic components in radioactive environments makes it a practical necessity to shield them and to allow easy access, but this can easily be justified by the mass of the large transporters concerned [see Figures 1.7.2.(1) and 1.7.2.(2)].

In late 1984, the name Remotex (that sounds like Purex, one of the widely used chemical reprocessing processes) was changed to Teletec (Feldmann, 1984) according to the generalizing development of the concept and to avoid confusion with the Remote Technology Corporation (Remotec, USA).

Japan and West Germany adopted the Teletec concept for their fuel reprocessing facilities and future plants. Power and Nuclear Fuel Development Corporation (PNC) in Japan has launched an important research and development programme using the three successive prototype servo manipulators from 1983 to 1986 (Kashihara, 1984; Kashihara *et al.*, 1983; Takanishi *et al.*, 1984; Maeda, 1982).

The side-elbow manipulator shown in Figure 4.1.2.(1) having been tested many times in the unilateral position servocontrol mode was tested in the bilateral mode in 1984. This delay accounts for the exclusion of the system from Chapters 1.4 and 1.5.

We know that this experimental arm is too slow, and there is a large lag between master movements and slave movements if the master arm is moved quickly. A computer-generated voice tells the operator to slow down (Kashihara, 1984). One pair of arms is carried at the lower end of a Cartesian bridge-mounted transporter bringing about two redundant motions (up and down and unit rotation about the vertical axis). This system was inspired by an earlier concept from Martin Marietta (1980) as the result of a contract from Oak Ridge. The 1980 system was equipped with two arms that moved in a horizontal plane and unit up and down motion. The Japanese arms employ a two segment polar planar displacement (without the effect of gravity) of the whole forearm and one extra motion that moves the wrist out of the horizontal

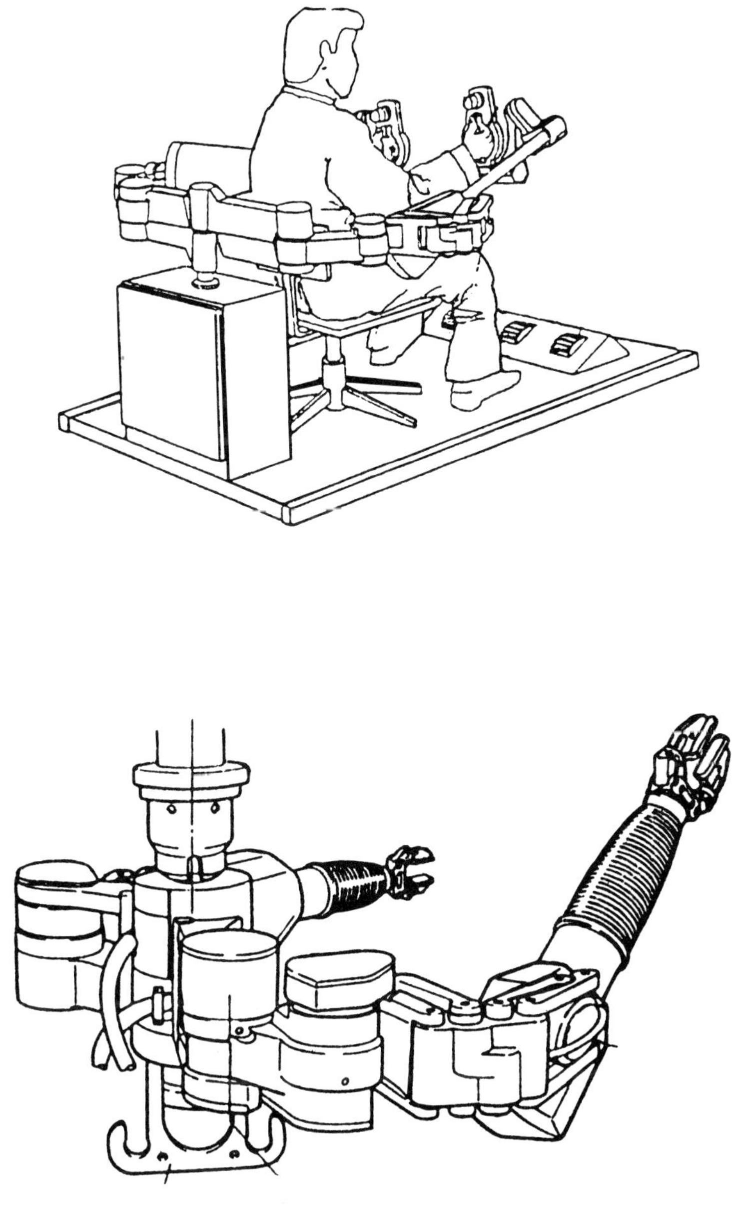

Figure 4.1.2.(1) *PNC manipulator. From PNC*

plane. In addition, the original wrist joint is equipped with a double active universal joint [see Figure 1.5.1.(1)].

In West Germany, DWK is running all full-scale process equipment in a test hall to test remote manipulation and the exchange of failed modules (small or large). It was verified that a simple bridge crane with a special impact wrench and/or an unilaterally powered manipulator was successful at changing modules. Nevertheless, by late 1984 no use for the more dextrous bilateral servo manipulators had been decided upon (Leister, 1984I, II).

Meanwhile, the Remote Technology Corporation (Remotec), founded by John White who elaborated the Remotex concept (now Teletec), is marketing, in addition to the engineering know how for reprocessing plants and their equipment, the new RM10 light duty unilaterally controlled servo manipulator [see Figure 1.4.5.(6)].

These efforts reflect the return of more complex (and expensive) manipulators that will allow substantial savings in the cost of maintenance of process equipment.

This Teletec concept comes down to extrapolating techniques already proven in hot cells and with accelerators, to the size of a real factory, with generalized use of television cameras and bilateral servo manipulators, associated with lifting devices (and high capacity arms for certain tasks) on large transporters.

It does, however, involve the use of a single large enclosure, and this precludes any contained transfer, which can be of great importance in nuclear teleoperation. Current thinking at the CEA is directed at a new approach, known as *teletrans*, which aims to integrate the two techniques in a more practical way, under the direction of Jean Vertut.

The importance of this application easily justifies all the research mentioned previously (and particularly that carried out in the ARA project), which will allow computers to be applied to the planning of maintenance and its real-time consequences, the automation of movement of large transporters with obstacle avoidance, the automatic adjustment of slave manipulator positioning by the transporter, the automatic tracking of television cameras and the various types of assistance for bilateral servo manipulators.

4.1.2.2 MAINTENANCE IN NUCLEAR REACTORS

The technical development of nuclear reactors has taken into account only the operations to be initially performed remotely, which are essentially the loading and unloading of fuel elements. It is interesting to compare different types of reactor from this point of view; the first reactors to use graphite as a moderator* had small fuel elements, which

*A moderator slows down the neutrons, allowing neutrons capable of sustaining the reaction to be derived from faster fission neutrons.

are easily handled in graphite channels. Heavy water* reactors are similar in this respect. Light Water Reactors (LWR) require enriched** fuel. This is the most widely used type of reactor, and it makes use of large fuel elements (4 m long) made up of 10 mm diameter needles.

Unlike the first types, in which loading is carried out during operation, handling takes place in a pool filled over the pressure vessel. When operation has been stopped, the pressure vessel is connected to the bottom of the pool, it is filled, and since the cover is opened under water the inside of the reactor vessel is directly accessible under water for handling with direct viewing (at a depth of 15 m). The pool is opened by a gate, allowing the fuel to be transferred to a second pool, intended for the storage of fuel prior to its transport to a reprocessing plant, or possibly final storage.

In this overview, the fuel progresses from small easily moved elements through channels, and gathered using standard blind mechanical methods, to a large mass which is put into place with considerable precision, under water. Comparable systems are found in FBR, but under sodium. In this context, teleoperation is strictly limited to handling remotely, first guided in the channels, up to work under liquid sodium at high temperatures. Some of the High Temperature Reactors (HTR) are an exception in which the ceramic fuel is coated in graphite to form pebbles which are extremely easy to handle in tubes. The designers of increasingly powerful and increasingly numerous light water reactors, only take into account, at the design stage, the fuel remote handling operations which take place about once every year. Apart from this, the remote handling and maintenance of highly radioactive objects has not been considered in these reactors, because the components concerned have been designed to have the same life span as the facility itself, (and as already mentioned, this does not take into account the unavoidable operations of dismantling and decommissioning), and any maintenance likely to be carried out with direct human intervention has been limited to areas in which the radiation levels are thought to be acceptable.

Practice has, unfortunately shown this concept of design based on normal industrial compromises to be inadequate. In particular, the basic phenomenon which complicates contact maintenance in light water nuclear power-stations is the presence in the *primary cooling circuit* (in which the water comes into contact with the fuel) of corrosion products in minute amount, and fuel canning activated inside the core, which results in a general deposit rich in ^{60}Co. The presence of this deposit limits the time of direct human access to the primary

*In heavy water, D_2O, deuterium replaces the hydrogen in ordinary (or light) water, H_2O.
**Natural uranium consists of mostly non-fissile ^{238}U with 0.6 per cent fissile ^{235}U. Enriched uranium is the level of ^{235}U that allows fission to be obtained in water.

cooling circuit when, after the reactor is shut down and the circuit is drained, there is work to be carried out on it.

In such a situation, where the initial design has not taken into account the need for remote maintenance, the problem of keeping the many reactors currently in use in a good state of repair becomes serious. The tasks required for maintenance of these reactors are as follows.

Inspection

The most fundamental inspection procedure is the periodic check, which takes the form of a visual inspection carried out routinely in all standard nuclear power-stations to ensure that everything is working normally. This type of inspection is useful to verify that there are no steam leaks, and that the state of the pumps and valves etc is normal. This check is, by definition, limited to those areas that can be reached when the power is on, and there is a policy of not carrying out checks very frequently or not at all in the more irradiated areas, even though it would be useful to be able to check these parts.

The second task of inspection is regulatory and intended to guarantee the safe operation of the facility. This basically concerns the checking of welded joints in all the vital parts of the primary cooling circuit (joints with the pressure vessel, welding in primary circuit tubes etc), and the condition of thousands of primary-secondary exchange tubes, where Pressurized Water Reactors (PWR) carry out the production of steam (steam generators). In Boiling Water Reactors (BWR), there are inspection valves and turbines which work with the radioactive steam.

This analysis, reduced to the simplest level, indicates that there is a need to inspect highly radioactive areas, and remote ultrasonic inspection of many welded joints of the reactor pressure vessel has been developed, using the In Service Inspection Machine. This device is introduced under water through the open cover of the reactor at the bottom of the loading pool. It is, in fact, a polar manipulator robot, programmed to move its ultrasonic sensors to well-defined spots, and to follow a series of welded joints that must be checked periodically. This system, therefore, runs a programmed operation rather than being an example of teleoperation.

The process of unloading the fuel under water, by teleoperation, is considered at the design stage, and it does not expose the personnel to very radioactive areas, except for operations to assemble to pool joint to the flange on the vessel, and the unscrewing of the cover which do require a certain amount of limited manual work (or the use of a special machine). In service inspection also involves only minimum exposure.

Inspection of steam generator tubes is carried out using a Heady current probe sent along the tubes by automatic machines, placed directly nearby the steam generators.

The amount of activated corrosion deposits in these tubes was not predicted, and the radiation levels are very high, but the inspection of the exchange tubes can be carried out without risk remotely.

Repair

These tubes will have to be repaired from time to time, which can involve either plugging them (thus putting them out of operation) or repair by sleeving.

All these operations around the steam generators' channel head, where there is a high level of radiation, can only be carried out for limited periods. These times must be shorter for work inside the channel head itself [see Figure 4.1.2.(2)], where an operator is restricted to two 10-min periods per year. This has caused, because of the radioactive deposits, a development from the initial simple techniques for manual work to sophisticated systems that limit the operator's exposure time.

This progression towards semi-remote work is similar to the automation process in other industries, but so far the result has only been to reduce exposure by only a factor of 2. Therefore, research and development into operations that can be entirely controlled from a distance has been encouraged.

Among the tasks commonly carried out close to the channel head of a steam generator (annual exposure of about five man-hours), is that of removing the insulation and opening the man way cover. In France, this is carried out by a special unscrewing machine, and another machine holds the cover (250 kg). Both machines are controlled by an operator who is present throughout. For work inside the channel head, a number of specialized machines have been developed, and this field is still expanding. In this way, it is possible to satisfy increased needs for maintenance and still reduce the exposure of the operators involved, to the small extent necessary, but if it would become necessary to reduce the annual radiation dose to which these workers are exposed by half in a short delay, the maintenance process would suffer.

A way of allowing operators to work in the channel heads has also been envisaged, not by using teleoperation to perform the maintenance tasks, but by first decontaminating remotely. The robot shown in Figure 1.4.3.(1) is intended for this task, and prefigures how it could be a maintenance robot which, instead of using a water sand-blasting nozzle would be equipped with specialized end effectors for the repair of steam generators. Westinghouse has produced the feasibility device, shown in Figure 4.1.2.(2). The hot points vary in different types of Light Water Reactor (LWR), but the basic problems are the same.

Emergency maintenance inside the reactor vessel or at the bottom of

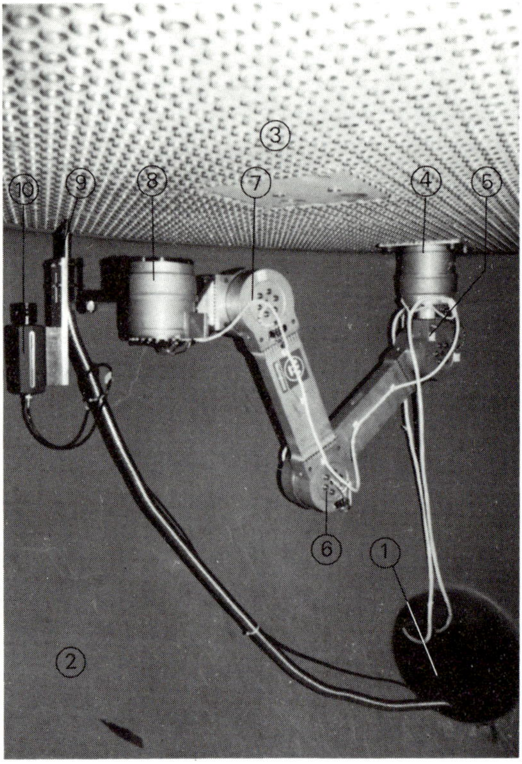

Figure 4.1.2.(2) *Experimental 'Rosa'* manipulator for the maintenance of tubes in nuclear steam generators (Westinghouse, USA, 1982). Internal view of a model of the lower part of the nuclear steam generator, known as the 'water box' (see Figure 1.4.3.(1)):* **1**, *man way with diameter of 40 cm, open during maintenance operations;* **2**, *smooth wall (sphere with diameter of 3 m;* **3**, *ceiling comprising 4000 exchange tubes allowing the heat of the radioactive water in the primary circuit to be transferred to generate steam with the water in the secondary circuit;* **4**, *base block, fixed into tubes on the tube sheet with expanding fingers. This block contains the electric actuator for rotation of the unit;* **5**, *identical actuator for the shoulder;* **6**, *elbow actuator;* **7**, *actuator for raising the wrist (pitch);* **8**, *yaw actuator;* **9**, *end effector for intervention on the tubes;* **10**, *camera integral with tool.*
Note: In an operational model, the actuators 7 and 8 would be smaller because of the small leverage; 4 and 6 would be of intermediate size.
Two operators carry the manipulator in four pieces, close to the steam generator, and assemble it. They fit the two outer rails parallel to the axis of the man way. They enter to position the television camera. Since the extremity 8 is fixed onto the rail-guided trolley, the trolley and the actuators can allow base 4 to be moved so as to be fixed in position using the television camera.
The main operator is, therefore, outside the reactor and the tools can be supplied by an aid to the manipulator, which extends 9 through the man way. By virtue of its shape, the arm can cover the entire tubular plate in two successive stations. Photo courtesy of Westinghouse
**Remotely Operated Service Arm*

the pools can take place remotely, working under water [see Figure 1.7.8.(1)], and it is common to employ divers, suitably protected, for work in slightly contaminated water.

A number of ideas have been developed, with the intention of launching research and development programmes to allow all these tasks to be carried out remotely, and to create *autonomous maintenance robots*, which are a halfway stage between self-adaptive autonomous robots and advanced teleoperators. This constitutes an application of CAT.

A machine of this type would be able to be taken under remote control by the operator in event of failure, and must be removed from the work area using pre-planned methods. It would certainly not lead to the facility being immobilized because of it ever breaking down.

Work concerning a robot for routine inspection was started in Japan a few years ago, more recently in the USA and Electricité de France has also made developments in this field. This work is complicated as a consequence of the fact that no provision was made for remote work to be carried out in the reactor *containment,** and since the means of access were designed for man the design of the autonomous vehicles is crucial. In particular, it is still very difficult to install a robot in a steam generator, with a vehicle to transport it from the entrance to the containment, and the only economy takes place in the last part of the journey. Hitachi is pursuing a line of study which aims to provide access to all points of the reactor containment, using vehicles equipped with both wheels and legs, similar to that shown in Figure 4.1.2.(3).

Fast reactors

Unlike water reactors based on circuit distribution, the most advanced French fast reactors have tended towards integrated construction in which the large vessel of liquid sodium contains, apart from the core of the reactor, an area for interim storage of the fuel, the pumps and the primary heat exchangers. This has allowed highly efficient repairs to be carried out on the primary-secondary exchangers and the pumps of the Phénix reactor, which have had only a minimum effect on the operation of the facility. The loop arrangement, adopted in the SNR reactor in West Germany, poses the same problems as are encountered in PWR. Our colleague Köhler has made a design of the vast teleoperator equipped hot cells which allow these circuits to be inspected and eventually repaired (Köhler, 1981, page 622).

The integrated reactor concept, adopted in the first commercial fast breeder power-station, Super-Phénix, which is nearing completion in France, has resulted in simplified transfer and teleoperation under

*The reactor pressure vessel pumps and steam generators are for safety reasons contained inside an airtight pressure-resistant enclosure called containment. Clean steam turbines and fuel storage pools are housed in the ***auxiliary building*** which is isolated from the other building.

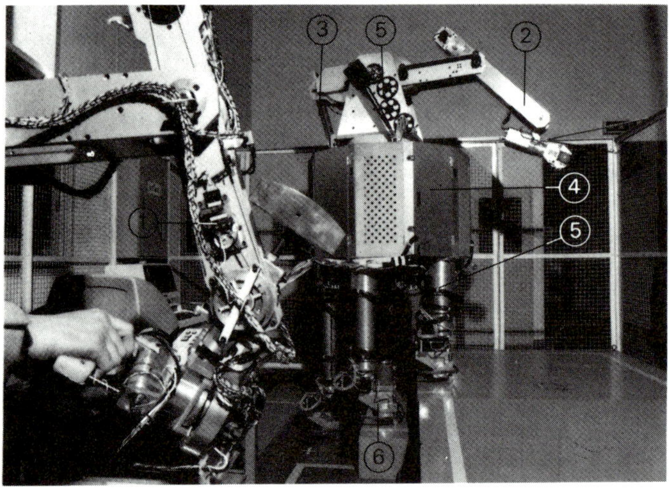

Figure 4.1.2.(3) *Experimental legged vehicle and a bilateral servo manipulator by Hitachi (Japan, 1982):* **1 and 2**, *master arm and slave arm [see Figure 1.5.0.(3)];* **3 and 3′**, *constant force springs used in combination with the approximate counterbalancing by the weight of the motors, and placed above the shoulder and elbow respectively;* **4** *body of the hybrid vehicle with five legs and five wheels;* **5**, *telescopic leg;* **6**, *orientable wheel support allowing rectilinear displacement in all directions and pivoting about the centre of the vehicle. It is equipped with one horizontal displacement in the plane of the wheel which at the limit of a step in a staircase allows weight to be placed on the next without the whole vehicle moving forward. This is often impossible while disengaging a single wheel. The vehicle is controlled by sequential commands relating to direction and distance. Sensors ensure that it remains vertical and that the vehicle stops when it encounters obstacles. Climbing stairs was made semi-automatic at the end of 1983. Photo courtesy of Hitachi*

sodium, carried out from above, thus simplifying all the common operations in the desired way, including that of final decommissioning, which was taken into account at the design stage. This reactor has a very large capacity vessel containing 3,500 tonnes of liquid sodium, surrounded by a safety vessel. During shut down, the space between the two tanks is lowered at a temperature of 180°C, and in this environment there are several kilometres of welded structures to be remotely inspected. The Machine for Inspection of Fast Reactors (MIR) has been designed for this purpose, and is an advanced teleoperator. It is based on the Mérite system [see Figure 1.6.7.(3)], and the existence of a vehicle providing access to this space, which is normally inaccessible to operators, provides a new outlook on the rectification of abnormal and particularly accidental situations.

It should be noted that in the specific case of the fast neutron reactor of this type, in the ultimate case of the reactor being cooled down, the solid sodium would constitute, after rapid radioactive decay, a shielding allowing direct human access, on condition that it is not exposed to oxygen, if any intervention into the mass of sodium is necessary.

Nuclear fusion reactors

This unlimited source of energy will, it is hoped, become widely available in the next century, and has led to extensive international cooperation. From the start of study into the first feasibility reactors and the conceptual design of the first industrial scale reactors, the question of their maintenance using teleoperation has been an issue. This is the reverse situation to that of currently operational reactors, the nuclear reprocessing industry and the underwater oil industry.

To understand this difference, it is important to be aware of the structural differences between the various facilities involved. In a reactor, harmless material is introduced (the new fuel) and radioactivity is created. There remains a relatively low level of radioactivity due to the activation of internal structures, once the irradiated fuel has been removed (although this level is generally underestimated at the design stage, and has imposed the conditions of semi-remote maintenance mentioned above).

In reprocessing, considerable levels of radioactivity are involved in the form of the fuel. This is introduced, processed and evacuated in the form of waste and purified products. Only a residual level of radioactivity remains in the facility (although higher than that in reactors), in the form of contamination in confined areas. Here a more extensive use of teleoperation should have been foreseen. The Teletec concept was designed to re-establish remote operation in this field.

In the case of fusion, the energy process itself does not generate great amounts of radioactivity, apart from that of tritium, which is necessary to the reaction, and is produced in slightly larger quantities than are consumed in the reactor. The intense activation of the structure of the reactor itself causes considerable difficulties, since the components involved are complex, massive and made of noble metals.

In addition, in contact with the vacuum chamber (which is either toroidal or cylindrical, depending on whether the system is the Tokomak or Tandem-Mirror), there are magnetic coils which should use super-conducting in the first industrial systems, and must naturally be protected from radiation by the nature of the organic insulating materials. There is, therefore, an initial shielding layer which can either be increased considerably so as to offer human access for direct maintenance (*contact maintenance*), or limited to the minimum protection, in which case it must be initially decided that fully remote maintenance is carried out. In addition, unlike graphite reactors or those which function under liquids, access to the inside of the vacuum chamber when the reactor is shut down poses certain complex problems. The welded stainless steel walls must be cut open to provide a limited opening into an ambient internal radiation of the order of 10^6 rad per hour, which is roughly the same as the highest levels in reprocessing plants. This combination of complexity and radiation levels has led to some large international projects (INTOR AND NET)* adopting remote techniques for all maintenance inside the torus, and most maintenance outside. Nevertheless, human access must be kept as a possibility whenever possible and profitable. The current plans for these facilities involve large areas and masses (dismantling a section of the vacuum chamber of the Tokomak would require moving equipment weighing about 300 tonnes to the nearest millimetre), automatic welding and conformity to the specifications of ultravacuum.

Considered strictly from the point of view of radiation, this situation is similar to that of the vast hall in the Teletec concept, with the distinction that the tasks to be executed would be far more delicate, involving many electrical components and considerable numbers of tubes. In opposition to this concept would be the presence of radioactive dust and tritium, which makes it necessary to try and reduce the contaminated volume. A new approach, which is unlike the overall contaminated volume in Teletec, which is to be called teletrans, takes into account the vital link between transfer under containment and advanced teleoperation.

It is envisaged that the removal of the various parts of a sector involved in a Teletec approach could be carried out in Contained Transfer Units (CTU), themselves containing a means of teleoperation

*International Torus with USSR, Japan and Europe (INTOR) and Next European Torus (NET).

added to the external vehicle-mounted systems already discussed in more detail [see Figure 1.7.2.(3)].

Unlike the components of other reactors, those from a fusion reactor must be removed intact from the reactor containment building, and then subjected to complex processes in a large hot cell situated in the ancillary building. The diffused tritium must be removed from the components by baking for repair, or the waste cut down for storage, if appropriate. In this way, for the normal operation of a fusion reactor, three types of teleoperation used in the nuclear industry are combined: exploration and inspection, maintenance, and dismantling without waiting for the facility itself to be dismantled.

This illustrates the importance of these reactor programmes for advanced nuclear teleoperation (and all it involves in common between fusion and reprocessing).

4.1.3 Decommissioning and dismantling nuclear facilities

Unlike fusion, which is directed solely towards future applications, decommissioning should take into account the facilities not yet planned for dismantling and/or those which have already ceased to operate. Keeping the latter in a condition of safety and protection is very costly over the years. Decommissioning techniques should be developed on these facilities which must, under any circumstances, be dismantled, for future application on installation that are still functional, and later still, for those in which decommissioning is planned at design stage. Every action of decommissioning currently undertaken is seen as a pilot project, and can help in the development and qualification of at least one prototype advanced teleoperator.

4.1.3.1 REACTOR DECOMMISSIONING

Each nuclear developed country has dismantled, or will soon have to dismantle one of its oldest reactors. These older facilities are almost all graphite reactors with low levels of radioactivity. The main problems with such an undertaking, relate to the handling and also the management of waste. Because of the decrease in activity over a period of time, the handling of active components from these first facilities does not present very great difficulties. For the first generation of relatively low power reactors all over the world, after the dismantling of the cooling circuit and controls located outside, there generally remains the graphite and/or internal parts in activated steel, inside the protective shielding. It should be noted that two teams working on reprocessing at industrial prototype level and one working on in service inspection are working on three old reactors, of which there remains the graphite in the well-sealed concrete protective shielding. This is the case with our colleagues from Oak Ridge, those at Marcoule, working with the MA23M manipulator, and those from Saclay who developed the MIR [see Figure 4.1.2.(3)].

In experimental pool reactors, underwater remote dismantling is easy and the same applies for the first water reactors to be treated in this way.

Graphite reactors

The dismantling of the G2 reactor, planned for some years hence is being studied in France. This facility had an electrical power of 250 megawatts, and operated under 15 atmospheres (15 bars) of CO_2, with the fuel loaded and unloaded during operation in horizontal channels. A conceptual design has been proposed for dismantling from the top, consisting of a reduction in the top thickness of the cylindrical concrete shielding to form a flat upper surface, and constructing an upper hot cell there, containing the means of Teletec-type remote load lifting and teleoperation.

This cell would include a transfer system, allowing the pieces of steel and graphite to be transferred to the outside. From the inside, an opening would gradually be made in the steel, combining the use of dextrous manipulators on transporters, heavy duty manipulators (acting as the left arm) and methods of load lifting adapted to the various large components. With these three types of machine, the dexterity function for precise cutting, the support function for heavy pieces, and the transport function for heavy debris to the transfer device would all be present. This general plan is based on the Teletec concept, inside the specially created hot cell, and the method could equally be applied to a fast reactor by adding a containment from oxygen.

In this situation, it is necessary to design and build additional contained transfer systems to carry out dismantling (see Section 4.1.3.2 in connection with reprocessing facilities).

Industrial light water reactor

In this context, most of the teleoperation takes place under water, for dismantling the internal parts using the ability established for operation and maintenance. There are, however, problems involved in handling the entire reactor pressure tanks, which may weigh between 330 and 430 tonnes depending on the type. This would have to be stored for some 30 years before the level of radiation decreased sufficiently to allow easy access. One solution to this, proposed by Bertin in France, consists of incorporating a storage space under the reactor containment building which would allow one or more reactor vessels to be lowered into the space, which would be well protected and, if necessary, another reactor of a similar type could be constructed on top. Since the dismantling of any contaminated primary circuit is carried out after decontamination, this would also leave a site containing an activated vessel which could be finally removed after 30 years; this plant provides a compromise between time and teleoperation.

Future fusion reactors

The current designs of these systems aim to integrate the means of Teletec or teletrans type teleoperation, which would allow intervention, repair and dismantling all to be carried out with specific tools, following a transformation of the hot cells to allow conditioning of waste from the decommissioning process.

4.1.3.2 DECOMMISSIONING OF REPROCESSING FACILITIES

Facilities of this sort which have either been or are to be soon dismantled are generally smaller than present reactors, and have given rise to some promising developments in France, following the policy described above, in which each site is used for a pilot project, involving the construction of a prototype teleoperator, the performance of which is evaluated during the operation. Jean Vertut has been deeply involved in the development of the Piade system [see Figure 1.7.2.(4)] for the decommissioning of caesium producing plant on the reprocessing site at the Hague. Elan II B extracted the caesium from concentrated nitric solutions of fission products. This plant illustrates the teletrans concept, since it consists of a teleoperator in juxtaposition with its transfer system, and since entry through the ceiling was possible without any new additions, the narrow existing passages were used. The system, which has been in use since the start of 1982, has been satisfactory within the normal limits of prototype operation and a fault which arose in a serious MA231 failure was easily repaired (Vertut *et al.*, 1984III).

A new problem arose concerning a large pilot plant for the reprocessing of fast reactor fuel (AT1), and is currently being analysed with a view to solving the problem when it arises on a large scale. The difficulty relates to the cutting up of relatively large pieces and removing waste, fully remotely. One method envisaged consists of constructing a transfer corridor prior to the removal of ceiling steel slabs, and allows direct removal of these pieces. The planned opening in the ceiling of the corridor would allow various points of access for the teleoperator which, like Piade, basically consists of a transfer hood which can contain the large transporter and one manipulator, and in this way constitutes a further progression in the teletrans concept.

The teleoperator has a greater reach than the Piade system, and cooperates with a bridge crane to lift large pieces, while another smaller bridge crane carries the heavy cutting tools.

The first bridge must transfer the large pieces of waste to an intermediate transfer container, which is itself transported by monorail along a special corridor. By combining the means of transport and transfer in this way with a remote cutting device, a new strategy for remote dismantling work can be developed, in which the compromise between the modifications of installation, waste production by remote cutting, and lifting of pieces cut to the largest possible size can be optimized by analysing the entire system (Clement *et al.*, 1984).

When decommissioning a facility which does not employ teleoperators in its everyday operations (or in accidents), there may be the need to reconstruct a facility with teleoperation capabilities. This will involve additional work to create a facility for waste conditioning and removal appropriate to decommissioning by having a much larger transfer capacity than the original plant had when in operation.

4.1.4 Emergency intervention

This concerns those very rare situations in which teleoperation may be the only possible solution, and may call for the development of special equipment. This may delay to a great extent the necessary technical interventions. This aspect may justify part of the exaggerated apprehension of the general public with regard to the nuclear risks. The Three Mile Island accident (TMI.2) took place in 1979, and the first entry into the containment building was not until two years later. Even now, the use of teleoperation is being envisaged to remove fuel in 1985-1986, but if the equipment had been available it would have already allowed useful entries to be made for the necessary intervention. It would have been entirely possible to develop a teleoperator for this between 1981 and 1983, if there had been a decision by the responsible engineers to do it in the time. The first exploration of the highly active basement started in November 1984 using a teleoperator vehicle developed at Carnegie-Mellon University, USA (Alexander, 1984). Jean Vertut had demonstrated the Oscar vehicle [see Figure 1.7.8.(1)] at TMI one year earlier.

Emergency intervention, for whatever reason involves an initial exploration to allow adequate observation by television camera and the mapping of radiation levels of the area, followed by work to allow the most important sources of radioactivity to be removed (liquids or solids) and, if necessary, the disassembling and removal of sources of medium level radioactivity, sufficient to allow repair to be carried out. In some cases this can be carried out by human intervention, once the area has been made safe, but other cases involve teleoperation for all the stages, whether for repair or decommissioning.

4.1.4.1 EMERGENCY INTERVENTION IN FACILITIES EQUIPPED WITH TELEOPERATION SYSTEMS

This would correspond to incidents in Teletec or teletrans type facilities, and would simply involve the introduction of the appropriate tools, and possibly the use of existing means for the dosimetric exploration or removal of strong sources of activity. The initial investment in teleoperators is partly recovered as an insurance and they greatly ease the recovery.

4.1.4.2 USE OF GROUND TELEOPERATORS IN PLACES NOT EQUIPPED WITH TELEOPERATION CAPABILITIES

General case

This concerns laboratories, which are accessible to vehicles moving over the ground, as well as the appropriate parts of nuclear reactors as reprocessing plants (although the limitations in these are severe).

The need for these techniques may also arise as a result of accidents endangering the environment, outside a building on a nuclear site, or beside a main road, in cases of transportation accidents. The extreme importance of having available intervention vehicles, which usually move over solid smooth surfaces but for emergencies, must sometimes have the capacity to negotiate obstacles and rough ground, becomes clear when incidents of this sort are considered.

A number of systems capable of this have already been considered in Chapter 1.6, including the vehicle-mounted manipulator by PaR, called Tomi in France [see Figure 1.6.4.(3)], the offroad caterpillar-tracked vehicle MF2 with radio control and television feedback [see Figure 1.6.4.(5)], the more agile devices, such as Mérite [see Figure 1.6.7.(3)], the MF3 [see Figure 1.6.7.(10)] or the hexapod devices derived from the Columbus hexapod [see Figures 1.7.9.1.(3) and 1.7.9.1.(8)], Odex [see Figure 1.7.9.(5)] and the device with five wheeled telescopic legs by Hitachi [see Figure 4.1.2.(3)] (Ozaki *et al.*, 1983), and more recently test teleoperators such as the Mitsubishi (Matsubara *et al.*, 1984) or the MITI (Nakano *et al.*, 1983).

Two other vehicles, the very small Oscar system [see Figure 1.7.8.(1)] or the multitask high-mobility high-dexterity Virgule teleoperator [see Figure 1.7.8.(2)], require special mention. The name Virgule is derived from the French initials for radio-controlled intervention vehicle. It is used in laboratories and offroad. One important feature of its design is the high-mobility variable geometry chassis (Vertut and Guilbauld,

1972), which can run on a variety of surfaces. It is a precursor to the multiservice self-powered and/or autonomous robots currently studied in the French RAM programme. This project is an international collaboration in advanced robotics that began after the Versailles summit in 1982.

In the absence of serious study of the global problem, Jean Vertut suggests a modular concept of teleoperators (for use on hard ground), and shows the relationships between vehicle size, its mission, manipulator capacity and reach using an intermediate transporter.

Smaller vehicle, of a type similar to the Oscar, can be used to carry a television and a means of manipulation of reduced reach and capacity. For heavy duty manipulators (25 to 50 kg) with a 1 m reach, a larger platform that can pass through a standard door (width 80 cm, as in the PaR or Tomi vehicles) is required. A telescopic column is capable of carrying the 25 to 50 kg capacity arm, up to a height of 3 m and still be able to reach the ground. This also applies to unilateral 50 kg and bilateral 25 kg arms.

To increase the horizontal and vertical reach, a distinctly larger platform is needed. When the use of an articulated transporter becomes necessary for the envisaged reach, and as manipulative capacity increases, the size and/or mass of the platform is increased considerably, with the total mass of the manipulator (five times the capacity) and the reach. It is easy to place the Monitor system [see Figure 1.6.5.(1)], since the vehicle is really a platform with passive mobility, the Mantis teleoperator with its reduced size for contact with the ground intended for use in the corridors at CERN, or the machine shown in Figure 1.7.2.(3).

This strictly geometrical approach does not take account of any specializations, apart from the ability to move over reasonably hard ground. It would be entirely legitimate to consider other important properties, such as properties of the vehicle, column, articulated transporter and manipulator modules to be watertight and resistant to oxidation, to allow washing in cases of radioactive or chemical contamination, and also for operation under water at reasonable depths, as in nuclear pools or in harbours and close to the shore (at depths of 20 to 100 m). Watertight and airtight devices with this type of capacity can be used at depths of 300 m, simply by pressurizing them with a skin diver air supply regulator.

Clearly, this matrix approach is a preliminary response to the broad variety of needs by a low number of machines, having a reasonable versatility. This may prevent a refusal to use teleoperation, and/or to start too late to develop methods that are not available, applicable to the nuclear industry as well as other hostile environments with similar characteristics. Of course, these machines must be of the RAM type (multiservice autonomous robots) using advanced teleoperation and

Figure 4.1.4.(1) *TELEM radio-controlled teleoperator (Ateliers de Constructions Electriques, Charleroi, Belgium, 1984):* 1, *two SM229 manipulators [TOS-USA, see Figure 1.5.0.(1)(d)] with 10 kg capacity, mounted on* 2, *articulations allowing them to be stowed to reduce the width of the vehicle;* 3, *pivoting, telescopic column;* 4, *two 12 V batteries;* 5, *space for electronic components;* 6, *block with two pivoting caterpillar tracks. In this photograph, the unit is shown without its television camera, which would require a separate transmission channel. Photo courtesy of ACEC*

aiming for an increasingly autonomous function along the lines of third generation robotics.

Being not yet in this next generation, the TELEM teleoperator shown in Figure 4.1.4.(1) was presented in the middle of 1984 by the Atelier de Construction Electriques de Charleroi (ACEC), Belgium (De Vresse *et al.*, 1984; Da Costa and Devresse, 1984; Flatau and Da Costa, 1984) and early test runs were announced for the beginning of 1985.

One pair of SM229 arms by Teleoperator Systems (USA) [see Figure 1.5.0.(1)(d)] is mounted on a column, carried by a Morfax-Marander vehicle (Köhler, 1981, p. 600). Special articulations on top of the column enable the slave arms to pivot to reduce the width in order to enter doorways.

The caterpillar-tracked front and rear pivoting blocks enable it to cross obstacles and to climb stairs.

The machine is powered by two lead batteries. These also feed the slave digital servo and the radio transmitters and receivers. There are two digital bilateral servos for the SM229; one was developed at Oak Ridge and the other in Belgium.

The electronics for this system is developed by ACEC in collaboration with Carl Flatau at TOS. TELEM is also expected to carry out routine inspections in nuclear power reactors.

Emergency interventions in the reactor building

At this level, the basic problem concerns the normal routes taken by operators, which involve stairs and lifts that are generally inaccessible following an accident, and therefore vehicles that climb stairs and enter narrow corridors are required. Such were the criteria for the development of the Virgule machine, on which work, unfortunately, has been halted since 1974, and the same considerations apply to the new Hitachi machine [see Figure 4.1.2.(3)]. They are also other Japanese approaches (Arai *et al.*, 1984). There is ample opportunity for innovation in this area.

It remains to be seen whether these developments can be justified in the context of normal maintenance.

Transport accidents

The extreme safety precautions taken in the transportation of radioactive materials (usually fuel) does not exclude the possibility of a transport accident requiring special intervention. Consider the special case of the nuclear arms that are continuously transported by plane and crashed into the sea at Palomares in 1966; the CURV machine [see Figure 1.1.3.(1)] recovered them and then several thousand drums of slightly contaminated earth were removed.

In the event of a ground accident, it may be necessary to approach a container carried by rail in conditions of slight radiation after a rail accident. Clearly, the main consideration is the deployment of equipment rather than remote control work proper.

It may also be of interest to use remotely controlled aircarft, such as helicopters or dirigibles. Finally, if it ever happened that a container leaked or came open, the levels of radiation would make it necessary to user remote control exclusively, with an offroad teleoperator capable of placing additional shielding around the container, or of closing it.

4.1.4.3 INTERVENTION IN THE ACTIVE PARTS OF A REACTOR

The pressure vessel of water reactors

Many special machines have been specially designed and operated to achieve remote repairs under water in water reactors. In many cases divers worked in these pools where special care was taken for radiation and contamination exposure.

The nuclear community is faced with unexpected problems relating to the unloading of severely damaged fuel at the TMI.2 station. The resources normally available for underwater work are likely to be inadequate, and the devices used will have to combine dexterity with an ability to assist heavy duty systems for the removal of heavy pieces of fuel. An approach similar to that used in the Piade applied under water would allow the very contaminated water, close to the core, to be contained from the rest of the pool so that waste could be removed in sealed drums. (This is the opinion of Jean Vertut.)

Emergency intervention in the primary circuit

The Mérite vehicle, shown in Figure 1.6.7.(3), or similar machines from Germany and Japan, designed solely for exploration or inspection of the circuit under normal conditions, when the reactor is stopped, become an indispensable tool for the more difficult work on the primary circuit, whether carried out under water or in dry conditions. There are particular applications in which the vehicles must work in opposition to gravity, with support from the surrounding walls. One example is the exploration and removal of fuel fragments from the upper areas of the TMI.2 steam generators.

The pressure vessel of graphite reactors

Damage due to the weakening of structures and their corrosion by CO_2 has led, over the last 20 years, to a need for delicate repairs involving the fitting of new structures, and welding of joints etc, and for this, teleoperators capable of entering the normal fuel loading corridors must be developed. Such devices have been developed, in particular in Great Britain, for repairs inside the pressurized chamber of graphite reactors. These are manipulators mounted on linear transporters [see Figures 1.4.1.(7), and 1.6.3.(3)]. Figure 1.6.3.(2) shows the machine designed by Jean Vertut, but never developed, for the accident at the Saint Laurent I nuclear power-station in France in 1970.

Figure 4.1.4.(2) shows a French robot remote manipulator, tested since the middle of 1984, and planned for use in the Electricité de France (EDF) graphite-gas nuclear power-station, Chinon 3, in 1985 (Raynal and Closson, 1984).

Five manipulators have been produced by Hispano-Suiza, who will also undertake repairs.

The difficulty of using such systems, because of the very restricted internal space and the number of obstacles in the reactor, has led to programming. In addition, the reactor internals are deformed. This demands an accurate model and rehearsal using positional servocontrol based on a master arm. First, the manipulator is safely inserted into the reactor to take measurements by televisual pointing and laser ranging. This allows the model to be accurately adjusted, and a repair programme to be developed and tested. Then the robots are inserted for automatic execution.

This manipulator, like the Taylor arm, can carry out welding and install new parts. A second arm is used to supply the parts and tools as shown in Figure 4.1.4.(2).

During the remote manipulation (programming) phase, the master arm (called the puppet) and the slave are servocontrolled as follows: the operator exerts a force and torque at the wrist and this being measured at the tip, generates an equivalent speed and rotation on the slave side. The slave displacements are reproduced by the puppet (master) using unilateral slave to master positional servocontrol.

Force sensors in the slave end effector measure the reaction force and torque in the contact to be obtained during execution of the task. Proximity sensors with an algorithm for adaptation of the slave can correct the centimetric positional error to get contact. In order to avoid collisions, contact detector plates on the slave arm will activate a safety brake, if necessary.

This system is not bilateral, but achieves a slow mode of behaviour more or less equivalent to this because of the self-adaptability of the slave, provided by tactile sensors on the arm and instrumentation of the end effector.

This approach, which takes no account of the work carried out in advanced teleoperation and is greatly influenced by industrial robotics, involves real-time computer programming.

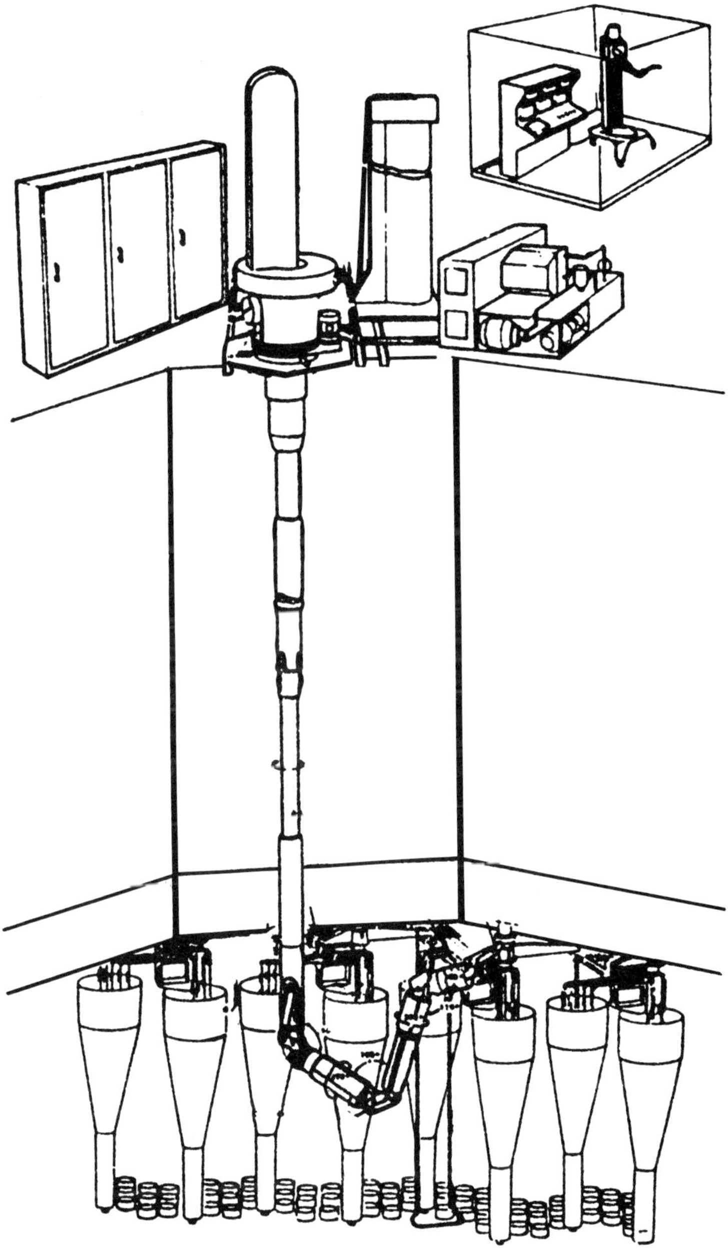

Figure 4.1.4.(2) *Computer-aided teleoperator (EDF-Hispano-Suiza, 1984, France). From Hispano-Suiza*

This remarkable achievement is to be admired, since it is not based on advanced robotics experience or university research. Its technological success is certainly linked to the practice of military equipment, which meets robotic techniques.

Conclusions

It is clear that the pressing need for teleoperation in the fields of inspection, maintenance, emergency intervention and decommissioning still encourages the nuclear power industry to develop advanced teleoperation techniques.

The modular approach using vehicles and transporters, both large and small, and a row of dextrous manipulators, allows the most varied combination of teleoperation systems to be assembled for work on hard ground.

This approach can be extended in the following cases:

1. hard ground with isolated obstacles, similar to the conditions for the use of Virgule, MF3, Mérite and legged vehicles;
2. an offroad situation, that is a slope or soil with low cohesion in which conditions the capacity of active suspension vehicles such as MF3, Mérite and legged vehicles are applicable;
3. when the terrain is very poor because of either the quantity of obstacles or the poor condition (25 per cent of all land is inaccessible to conventional vehicles), legged locomotion is the only solution;
4. when displacement is necessary inside facilities and equipment, with the particular problems of having to climb up walls and ladders, even legged locomotion may not be adequate;
5. for displacement on fluid interface and into a fluid, there are all the known methods of floating on the surface or under water, as well as airborne systems, with new aspects appearing linked to the absence of fixity in these vehicles and to problems of power supply.

This rapid modular approach neglects the important role played by transport and transfer systems in nuclear robotics, and the general study of this problem merits a more thorough treatment than can be provided here.

It should be noted that these maintenance and intervention robots (or advanced teleoperators) relate to the facility in a very different way to that of industrial robots associated with flexible manufacturing processes. The latter pay for themselves in the course of their entire working life, whereas the former can pay for themselves in a single operation, because the cost of immobilizing the facility would pay for these machines after one of two days production.

This is true of all intervention devices in accidental situations and in exceptional maintenance. This applies in nuclear maintenance, as well as for underwater oil, and in outer space applications.

A similar situation arises in the application of teleoperation in the fight against natural and artificial disasters, particularly fires or terrorism.

This seems to have been adopted, in Japan, to judge from the title 'robots for critical work' of the large scale research and development programme, started in 1983. This is based on work which is critical because of its difficulty, importance, duration or risk, which shows obvious similarities to this subject.

This programme has been named Jupiter, and relies on future international cooperation in advanced robotics, the French branch of which is called Autonomous Multiscience Robots (RAM).

This will be concerned with the new concept of service robotics which includes all these elements and, in addition, the devices linked to the industries sector not directly concerned with production.

Chapter 4.2

Underwater applications

Underwater devices, since they are normally unable to leave harbours directly, involve a transfer function which is distinct from transport. The exception to this is the French Argyronète interrupted project, now being developed again by COMEX [see Figure 4.2.1.(1)].

Underwater devices are transported by boat, or in some cases through the air, but all must be transferred underwater. This constitutes one of the most significant technical problems if intervention is to be possible in all weather conditions. Underwater teleoperation (in common with work in outer space to be explained in Chapter 4.3) is primarily intended for exploration, which is basically the same as inspection in industrial applications. Survey is necessary before maintenance, and, when accidents occur or for safety reasons, exploration can constitute the first part of an emergency industrial intervention. This implies that nuclear and underwater hostile situation systems, which both justify the use of teleoperation, have basic missions in common and similar strategies for deployment.

4.2.1 Manned submarines

The use of submarines, like space exploration and entry into hostile environments in general, is the subject of extensive debate over the presence or not of man in the hostile situation. Limited attention is given here to manned submarine devices, although these often form the basis for remotely controlled systems, as shown in Figure 4.2.1.(1), which illustrates the continuation of the Argyronète project by COMEX. The first underwater remote manipulators were intended only for work outside the human survival capsule. Cooperation between the human work and that of the remote manipulator is evident in this example of the submarine which 'spits out' its divers, and has sufficient power autonomy to avoid surfacing and launching in very heavy weather.

There are numerous other manned tethered spheres which, without being diving bells, are suspended from an umbilical connection tube, like ARMS by Oceaneering used in particular with the AS30 or the Diver

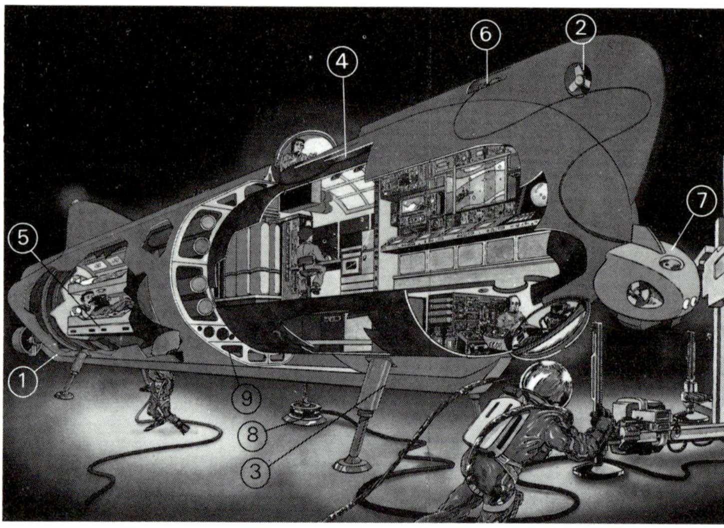

Figure 4.2.1.(1) *Projected submarine for experimental compound intervention (Argyronète, COMEX,* France). This project launched in the 1960s by the IFP** and the IFREMER*** was stopped late in development. It has now been taken up by COMEX, and an initial demonstration is planned. It applies the teletrans concept for underwater application, for transfer with operators on the spot. The Argyronète leaves the harbour directly and dives for its work area. This solves the problems of transfer across the surface at the price of considerable energetic autonomy:* 1, *main rear propellers;* 2, *manoeuvre propellers;* 3, *telescopic feet which rest on the sea bed;* 4, *cabin at atmospheric pressure, equipped with two large portholes;* 5, *decompression chamber for divers;* 6, *launching platform for remotely operated vehicle (ROV);* 7, *winding mechanism for tether of the tools;* 8, *engine room with Stirling or diesel motors in closed circuit at near atmospheric pressure. From COMEX*
**Compagnie Maritime et d'Expertise*
***Institut Français du Pétrole*
****Institut Français de Recherche et d'Exploitation des Mers (IFREMER), earlier known as Centre National d'Exploitation des Océans*

Equivalent Manipulator System (DEMS) arm with bilateral servocontrol, by General Electric. Following the lead of Graham Hawkes, submersibles have tended to be reduced in size to the extreme example WASP, which resembles the articulated diving suit JIM shown in Figure 1.3.10.(2). This system retains the arms and provides the operator with a means of moving around on the end of his tether, using between four and six propulsion units (*National Geographic Magazine*, 1983). In this situation, the operator is in a perfect situation for observation but his dexterity is severely reduced. Mantis, a very small manned submarine, equipped with electric manipulators, is currently in use for depths of more than 1,000 m. The Deep Rover, a small tracked vehicle roving on the sea bed, was also designed by Graham Hawkes.

Small research submarines are equipped with simplified manipulators intended for biological or geological sampling, and the small submarines associated with petrochemicals industry have slightly more sophisticated manipulators. The idea of a dextrous low capacity right arm associated with a strong left arm is widespread, and a typical example is the Work System Package (WSP) developed by the NOSC (Wernli, 1979), in common with several deep submergence American exploration submarines and the RUWS (lost at sea).

For applications associated with underwater petrochemical maintenance, the precision of positioning required may involve the capsule being attached to descent cables (guiding cables) or, as in space capsules, one or more arms may be used to attach the submarine to the structure, with one higher dexterity arm reserved for execution of the task, once the intervention module is safely fixed in place. There are about seven bilateral servo high dexterity force reflecting arms (DEMS or AS30) in use. Few users have already admitted the benefit of this faster execution of tasks. Manipulator dexterity in an underwater device must be in proportion to its agility (unless the high dexterity arm allows agility to be achieved indirectly, by attaching itself suitably to the object of the task).

The following anecdote illustrates the prevailing attitudes in this field. The CYANA diving saucer was equipped with a sampling shovel, shaped like a double scoop, and the diver, well known by his nickname 'Canoé' diver from the days when he was a member of Cousteau's team, had to adjust the position of the arm and then take the biological or geological samples by moving the little submarine itself, because it was capable of greater agility than its arm. When Canoé visited Vertut's laboratory he was astounded by the dexterity of the manipulators he found there. This proves that there is still a great deal of work to be done in equipping most existing manned or Remotely Operated underwater Vehicles (ROV) with bilateral servo manipulators.

4.2.2 Cable-controlled devices (Busby, 1981)

Without entering into the debate on whether underwater vehicles should be manned or not, and on the future use of divers generally, it is true to say that there is a progressive increase in the number of cable-controlled devices used in this field. Figure 4.2.2.(1) provided by the French Institute for Research on Sea Exploitation (IFREMER) (new name of the well known CNEXO) compares manned submarines for use at depths greater than 500 m and similar remotely operated cable-controlled systems, in terms of their numbers throughout the world. Numbers for 1976 in a survey conducted in the USA showed that there were 25 manned submersibles for use at depths greater than 500 m out of a total of 50, and that, at the same time, there were 45 ROV systems controlled by cable. Sheridan and Verplank (1978), who group active cable-controlled vehicles and towed systems together, compared the masses involved, and showed that out of the seven cable-controlled devices capable of descending to 6,000 m, six were passive towed systems for taking photographs or measurements, and only one was an active working system, the Remote Unmanned Working System (RUWS), which was lost soon after, as compared to two manned submarines, the Trieste bathyscaphe (USA) and the Archimede bathyscaphe (France), which has not been operational for several years. The mass of the RUWS was about 2 tonnes, as opposed to some 50 tonnes for the bathyscaphes, but this mass is only that of the slave unit itself, the transfer mechanism of which, as in Eric II [see Figure 1.6.1.(1)], includes a heavy unit for keeping the cable taut, which means that the mass that actually descends to 6,000 m is very much greater than that shown, but nonetheless it is only increased by a factor of 5 or 6. This mass is still less than the mass of the recent SM97 Punch 6,000 m manned submarine launched in late 1984.

Among the deep sea cable-controlled systems, there is an increasing number of relatively passive photographic towed devices, used for the surveying of the sea bed, for finding wrecks and sites of geological interest (underwater heat sources, polymetallic nodule fields). It is interesting to note that the Eric II project is, in fact, the PARC-Eric project, and the PARC system shares the same surface support, the same methods for launching and anti-crush suspension, the same cable and the same transmission.

The PARC is entrusted with the ambitious mission of photographing 100 m^2 of the sea bed, at depths of 6,000 m to help in preparing for the Eric II missions, and has reached the trial stage. There are more devices of a similar, but more modest, type for photographing the sea bed on a smaller scale. One deep tow system, after seven months of research in the Atlantic Ocean, was able to locate the wreck of a nuclear submarine lost in an approximately known area of some 100 nautical miles, with

	manned submarines for more than 500 m submergence	remotely operated vehicles (ROV)* (cable controlled)		autonomous vehicles	deep towed systems
USA	10	17	114	4	6
USSR	1	1	1	1	1
Japan	1	2	3	/	1
Germany	/	2	4	/	1
Great Britain	1	5	9	/	
Canada	/	3	41	1	
France	2 2	4	3	1**	1 2 in production

Figure 4.2.2.(1) *Comparison of numbers of civil submarines manufactured in the major industrialized countries*
**First column: number of different models*
Second column: total number of units produced, although certain types have been manufactured in large quantities
***The autonomous vehicle 'Epaulard' is the only one in this category to be fully operational*

the magnetometer and camera detecting it at exactly the same moment. These devices herald the introduction of robotics into underwater applications, starting with observation, followed by manipulation and, finally, the handling of heavy loads.

The main problems with these deep submergence systems relate to the cables both as a means of mechanical support and signal transmission. A coaxial cable which, of necessity, carries high voltages, must be capable of supporting several tonnes to depths of 6,000 m (requiring a length of some 9,000 m), also has to support its own weight, and must be based on advanced versions of current techniques, using synthetic fibres such as Kevlar.

The PARC-Eric II cable, which has a diameter of 30 mm, can transmit 100 kVA and simultaneously allow control and feedback signals to be modulated, including a single television channel. The future of this technique will be based on signal transmission using fibre optics for which, until now, the problems of connectors have not been solved industrially for these applications. With less deep immersion, the tendency is to use systems with independent power sources and a very thin connecting cable for fibre optic communication. This is the method adopted in the NOSC project, which is

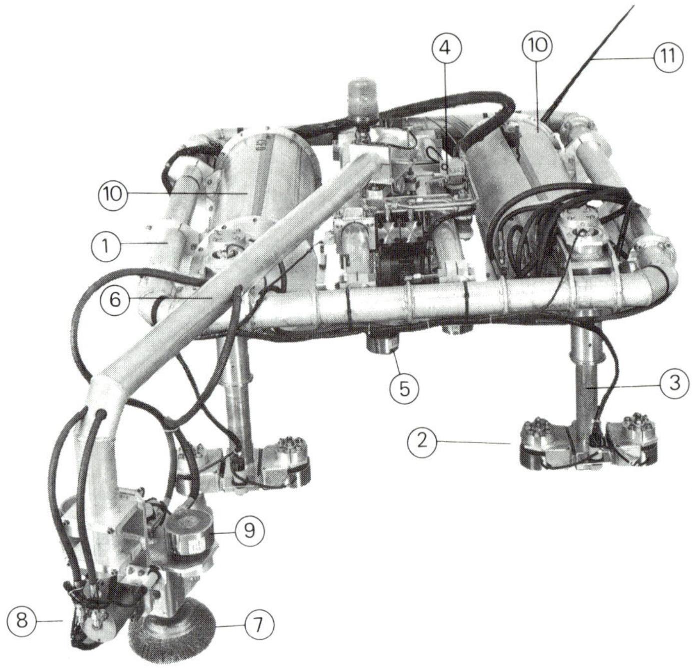

Figure 4.2.2.(2) *Marine robot: legged vehicle used to clean the hulls of ships (CNM,* France, 1983):* **1**, *tubular frame of the vehicle;* **2**, *feet equipped with double pads or electromagnets (on this picture);* **3**, *three telescopic legs powered by hydraulic actuators. The third is hidden behind trolley 4, as is the foot;* **4**, *trolley running on two rails;* **5**, *trolley feet;* **6**, *pivoting arm on the trolley;* **7**, *brush (or tool). This programmable teleoperator can hold a grinder, radiation source, high pressure water jet, paint spray-gun (used on dry hulls), television inspection camera etc.* **8**, *rotary actuator for moving the tool and foot* **9**; **10 and 10′**; *housing for the electronic components;* **11**, *cable for electrical and hydraulic supply, including control signal transmission by fibre optics. This robot can be programmed by teaching it the three DOF. The walking movement is achieved after moving the trolley forward, by fixing the arm and foot 9, as well as foot 5 by shortening the three feet 3, causing them to adhere to the surface then, after detaching and raising feet 2, advancing body 1 by a relative reverse of the trolley. (The transverse stability is ensured by a very slight release of feet 2.) Change of direction can be obtained by rotating the arm prior to feet 2 being placed. This process is identical to that used in the walking vehicles designed by Komatsu [see Figure 1.6.8.(5)]. Photo courtesy of Normed*

**Chantier du Nord et de la Méditerranée*

being developed with Professor Sheridan, using the supervisory control approach. This work has already entered the field of advanced teleoperation in the sense put forward in this volume (Sheridan and Verplank, 1978; Cowan *et al.*, 1980).

The most advanced, but heaviest, devices need an inertia platform to allow them to achieve precise navigation despite any currents, and positioning at a fixed point in relation to a structure to which they must be attached, or on which their manipulators will work.

The large population of cable systems for work at depths greater than 500 m includes observation systems as shown in Figure 1.6.7.(2) and many simple teleoperators which do not actually fall into the category of dextrous or advanced teleoperation systems.

The expected opening in underwater petrochemicals applications will spark off a technical development in the direction of advanced high agility and dexterity teleoperation. A recent development at Deep Ocean Technology by Graham Hawkes, the designer of the rigid diving suit JIM, WASP, Mantis and Deep Rover, is the Rig Rover or line-guided teleoperator, which almost constitutes a *mechanical diver*. The device descends on two cables, and is capable of a certain mobility by propulsion. The designer announced a force feedback arm with tactile sensors.

The system shown in Figure 4.2.2.(2) is, to the best of our knowledge, the only marine walking robot to be used industrially for climbing up and down walls, under water (or in air). It was developed by the Chantiers du Nord et de la Méditeranée for underwater work on the hulls of boats (eg cleaning, painting), and is the result of work by P. Kroczinski and the newly formed French company INRO for computer control and fibre optic signal transmission. It is similar to the MAM vehicle [see Figure 1.6.7.(4)], for locomotion on vertical surfaces, and to the octoped machine by Komatsu [see Figure 1.6.8.(5)], in its method of locomotion. Its main characteristic, apart from the use of fibre optic communication, is the use of the tool-carrying arm as a leg when the three main feet are in contact with a surface (Kroczinski, 1984).

4.2.3 Free-swimming underwater vehicles

To the best of our knowledge, the Epaulard machine [see Figure 4.2.3.(1)] by CNEXO is the only free-swimming vehicle currently fully operational to 6,000 m underwater, at which depth it has carried out several systematic photographic assignments on polymetallic nodule fields. Nevertheless, there are six other experimental machines in the same category, in particular, the exploration machine the Advanced Unmanned Search System (AUSS) (Estabrook, 1982), which is being developed by the American navy.

The Epaulard machine does not employ an advanced robotic approach but constitutes an exceptional development because of its hydrodynamics, its autonomy and its ultrasonic remote control. Its only robotic characteristic is that it automatically stops in front of any detected obstacle and sends a signal to the surface. The remote control, which requires a 4 s delay for submersion at 6,000 m, is thus coded with great security and commands are not executed before being repeated on the surface for even greater security. The machine is held at a fixed distance from the sea bed by a *guide rope*,* which is also used for the ascent. The Epaulard is shown in Figure 4.2.3.(1) under the gantry which ensures its launch and recovery (and one which is capable of much bigger and heavier machines). The hull is made of a moulded buoyant material, the controls and the propeller are at the rear and the camera is as the front. A second, much larger, machine carrying lateral sonar is currently under consideration.

It can be seen in this instance that outer space teleoperation is not the only field to suffer the consequences of transmission delay.

An experimental free-swimming vehicle which was designed under the direction of Jean Vertut is nearing completion; this PLA2 vehicle is shown in Figure 1.6.7.(5) and, as a result of the exploration done by Epaulard, will carry out basic technological experiments on the collection of polymetallic nodules (see Section 4.2.5). This vehicle uses, to an even greater extent, the structural and buoyant properties of moulded buoyant materials. It remains apparently a first generation programmed robot but has several functions for obstacle avoidance and for safe adaptation to unexpected events. It is the forerunner of a more robotic generation.

These machines, rare until now, open the way to a growing number of fairly self-adaptive autonomous machines which will really constitute the fleet of second or third generation robots intended for underwater work. Strictly speaking, they are remotely controlled vehicles rather than teleoperation tools, but the introduction of advanced remote manipulators possessing a certain level of control autonomy, controlled after a transmission delay using computers, will take place in the next few years as a result of the current work of NOSC and Sheridan (USA) and in the ARA and RAM projects (France).

4.2.4 Inspection, maintenance and construction of underwater facilities

4.2.4.1 SINKING AND EMBEDDING OF CABLES AND PIPELINES

The first of the underwater tasks developed since the last century was the laying of telephone cables, and the dramatic stories of cables lost

*Cable or heavy chain suspended section that weighs on the machine, with the section resting on the bottom lightening it. This is used for the landing of balloons.

Figure 4.2.3.(1) *Epaulard free-swimming vehicle (IFREMER, operational since 1980, France). This vehicle built by ECA, communicates with the surface by coded ultrasonic signals. It is used for photographic missions to depths of 6,000 m submergence, particularly of polymetallic nodule fields, and in 1983 of polymetallic deposits produced by hydrothermal springs. It moves along at a constant height above the sea bed (2 to 3 m), using a guide rope with slight drag partially lying on the ocean floor (flat chain made of cast iron) which ensures descent and is released for surfacing:* **1**, *hull made from large moulded blocks in a buoyant material (syntactic foam: hollow microspheres of glass in epoxy resin) fixed onto an aluminium skeleton. The form is adapted for vertical descent under the excess weight of the guide rope.* **2**, *securing system for launching and retrieval;* **3**, *ultrasonic signal transmitter-receiver;* **4**, *miniature radio transmitter and flash for surface location;* **5**, *propeller;* **6**, *compass servoed steering (the course to be followed is transmitted ultrasonically);* **7**, *powerful flash for photographs, hidden by the cradle;* **8**, *battery housing (can be jettisoned by remote control if normal ascent fails);* **9**, *high autonomy camera for photographing the sea bed, recording time and absolute depth (measured with an accurate manometer);* **10**, *sonar for detection of frontal obstacles. Detection brings about automatic stopping and a short reverse movement, signalling this to the surface. The operator then devises a new course so as to avoid the obstacle. The vehicle is located from the surface using short base detectors on the ship or from long base ultrasonic transponders on the sea bed by triangulation. Photo courtesy of IFREMER*

Figure 4.2.4.(1) *Intervention and maintenance remote manipulator (TIM, ELF Aquitaine, built by ACB, France, 1980). This teleoperator, built by ACB, was used experimentally on an underwater oil production site, with several well heads which it served by moving about on rails*
(a) *Rear view*
(b) *View from the left side*
1, *telescopic pivoting boom with hoist and electromagnetic grapple (four movements plus grasping);* 2, *shoulder of two manipulator arms on telescopic polygon columns to resist transverse rotation;* 3, *upper arm with two shoulder movements, rotation about itself, telescopic elongation and elbow movement;* 4, *forearm with telescopic elongation, gripper rotation and grasping. This arm with eight independent movements plus grasping is redundant for two usual reasons: redundancy of control (extension of the gripper) and avoidance of obstacles.* 5, *guide roller on the rails giving end of track command;* **6**, *bogey with two motor rollers;* 7, *hydraulic connectors for the entire left arm allowing removal of the whole subassembly on board the surface barge;* **8**, *container housing hydraulic control electrovalves (open here for servicing the machine);* 9, *large guard cradle, folding up to the vertical on the right-hand side, used during launching and retrieval through the central well on the barge;* **10** and 11, *cylinders for control of folding back;* 12, *similar guard bar on the left-hand side. Another,* 12′, *protects the other extremity of the trolley* 13. *The operator controls 18 two-directional switches placed on a diagram of the two arms, five on the diagram of the crane, and one switch controlling the translation of the trolley. All of these are on-off velocity controls. The operator also uses four television monitors. Photos courtesy of Jean Vertut*

1
2
4
3
10
11
12
eap
13
5
6
7
8
a

1
4
3
2
12
12
5
13
2
b

and found, and the difficulties in repairing them are well known. Since the time when the first telephone link between Europe and America was laid, techniques have greatly developed; today telecommunications cables have been joined, over great distances, by power cables. All these cables are sunk into the sea bed. These tasks are carried out by special machines that are trawled and controlled remotely from the surface and just fall within the range of this subject. Increasingly advanced tasks must take place on the sea bed, whether the sinking of underwater cables, oil or gas pipelines, or their embedding in rock when protection in light soil is impossible. A self-propelled embedding machine has been introduced. This French vehicle called the self-propelled embedding vehicle for the protection of pipelines (VELPO) forms part of the approach of advanced teleoperation in the sense that its loading hopper sensors ensure control of the positioning of the boat so that the bottom of the down pipe remains above the funnel while the pilot who, using ultrasonic sensors, sees the outline of the land in front of and behind the vehicle, ensures its remote guiding to maintain centering and to negotiate various obstacles. The design and use of this vehicle, in which Jean Vertut's team took part, is part of a venture between the CEA, the DORIS and OTP companies and the North Mediterranean shipyards. The tests at sea proved successful.

4.2.4.2 REPAIRING PIPELINES

This is the most pressing theme and the cost of laying deep pipelines largely justifies their repair as opposed to their replacement. Several approaches have been envisaged using, until now, either divers or men in atmospheric pressure bells, and recently in the French COMEX project an advanced teleoperated welding robot equipped with a three-dimensional vision system for following the exact profile of the seam during welding has been introduced.

This teleoperated intervention robot justifies its cost in just one successful operation. The discontinuous profit making capacity of advanced teleoperation systems applied to critical repairs is clearly illustrated.

4.2.4.3 MAINTENANCE ON UNDERSEA OIL PRODUCTION WELL HEADS

Today, machines are tested with a view to replacing divers for maintenance (or even installation) intervention in well heads in deep water oil production. This is one example of the debate between manual and remotely manned systems with human control where tasks cannot be directly automated. Another debate concerns manned or automatic systems. It is at the centre of the zone shared by advanced teleoperation and advanced robotics, which aims to carry out tasks that are increasingly more difficult to define, in an increasingly autonomous way, as shown in Figures 1.2.6.(3) and 1.2.6.(4).

The Rig Rover, developed by Graham Hawkes, is under development for oil production facilities, and Figure 4.2.4.(1) shows the French Telemanipulator for Intervention and Maintenance (TIM) tested by ELF and built by ACB. This teleoperator is fitted to a trolley that runs along the length of the facility and carries several hydraulic unilateral remote manipulators and a corresponding number of television cameras. It illustrates the difficulty of moving from specialized to multipurpose equipment; the absence of any adaptation of the well head necessitates the use of dextrous manipulators with advanced controls, which both designers and users have been reluctant to adopt; as a result there are many special devices and independent arm movements (that avoid computer coordination). The minimum of adaptation to teleoperation in the well heads would have contributed further to total optimization (equally favourable to intervention by diver).

This is another example of the debate between innovation and engineering which, due to the lack of opportunity for dialogue and objective demonstration of the techniques being developed, delays the adoption of techniques that are considered advanced by those who developed them and insufficiently demonstrated by those who cannot adopt them.

4.2.5 Undersea mining: polymetallic nodules

These beds contain 2.5 per cent copper and nickel, 30 per cent manganese and a small amount of cobalt. To use these resources it is necessary to collect potato-sized nodules from soft ground at a depth of 6,000 m. The production capacity is in the order of 1 to 3 million tonnes per year. Several approaches have been considered and some are the subject of small-scale feasibility tests. Three processes have been envisaged.

The first procedure, which is a direct operation using cables, is called the Continuous Line and Buckets (CLB) procedure. It was invented in Japan and perfected in France by CNEXO. By pulling up the cable at the front of one boat and letting it down again at the front of the second boat, a large section of the cable loop can remain on the sea bed. Buckets, attached to the main cable by 50 m lengths of line, travel on the sea bed and dredge. Finally, they are raised by moving only the main cables and by the slow forward movement of the two boats. The dynamic behaviour of this mechanical process was tested at the model stage. It has not been properly tested at sea following the failure of the test at sea using a single boat. This may imply strictly mechanical methods, but they are limited to rather small systems for reliable production, which may reduce the overall cost effectiveness of the operation. Problems of reliability have not been considered.

The second procedure, derived from undersea petrochemical techniques, uses a riser (vertical or approximately vertical pipe) for hydraulic lifting by mechanical pumping, or an *air lift*. The two methods have been tried on a small scale by international groups formed by American companies. The main problems lie in the pumping ability and in remotely controlling the dredge at the foot of the riser. This is a large heavy device, and the difficulty of raising it from the sea bed imposes the greatest reliability, which is difficult to guarantee. The problems in controlling this type of system are clear in view of the fact that the three-dimensional structure of the sea bed is more varied than is generally realized. The system's drawback is its very long response time to obstacles that are detected too close, or not thoroughly known in advance. Finally, the riser connection between the sea bed and the surface works in series, where any failure will stop the entire system.

An entirely robotic system has been envisaged by a team led by Vertut as the third procedure. Autonomous dredging shuttles work an underwater site. About 20 dredging shuttles each performing more than two cycles daily would allow an annual capacity of 3 million tonnes, and they could be inspected (and if necessary repaired) whenever they surfaced. This is a teleoperation system which falls into the category of advanced teleoperation, because of the high level of shuttle adaptation required. The PLA 2 vehicle, shown in Figure 1.6.7.(5), is a quarter-size representation of an industrial shuttle that is capable of lifting 250 tonnes of nodules per cycle. The total mass to effective load ratio is a merit factor of the greatest importance to its technological development. Figure 4.2.5.(1) is a conceptual drawing of a shuttle which could have a mass of between 1,000 and 2,000 tonnes. The GEMONOD group formed by the CEA and IFREMER is currently studying this project on a long-term basis, and an advanced version of the hydraulic riser. The experimentation of the PLA 2 will constitute a major step, in 1985 and 1986.

The problems of teleoperation for retrieving returning shuttles should be tackled by arranging underwater docking under a large surface platform requiring sensors able to operate in the ambient noise and ending with a shuttle to the surface, in case of any failure of the automation.

Another approach currently being considered would involve retrieval on the surface with relatively small cable-controlled vehicles, the mobility of which would allow easy docking, provide a power supply at the end of the mission, and direct control over the surface for entry into a door with an entrance lock of the support platform.

This solution, which may appear marginal to the object of the study as a whole, was included because it contains all the problems of advanced teleoperation in the widest sense of the term, and those of computer-aided intervention by operators in a process involving many

Figure 4.2.5.(1) *Conceptual design of an underwater nodule mining works, with autonomous shuttles (CEA-AFERNOD, France 1982*):* **1**, *surface support unit, here a semi-submersible platform that allows underwater docking;* **2**, *ore tanker transporting nodules or ballast;* **3 to 7**, *mining shuttle cycle. This cycle runs for about 10 h, and six shuttles make up the loop. 1 is finishing its delivery and preparation cycle (not visible), 3 is descending, 4 is touching down on the sea bed prior to dredging for about 1½ h, 5 is about to start surfacing, 6 is halfway up and 7 is on the point of reaching the platform. Four identical loops constitute the whole system, and work in areas sufficiently distant from each other to avoid any risk of collision. Two successive oblique passes will provide a sufficient return if the shuttles are not equipped with recognition sensors allowing dredging in rows. Figure 1.6.7.(5) shows the scaled down experimental vehicle PLA 2 that will be tested to provide technological data on the sea floor operation.* **8**, *Archimedean screw supporting propellers;* **9**, *silo for ballast or nodules;* **10**, *loading conveyors;* **11**, *return conveyors;* **12**, *independent modular dredgers including their own lifting belts feeding 11;* **13**, *propellers for movement in water, used particularly for another platform docking and maintaining the trajectory. The shuttles glide because of the effect of a slight weight for the descent, and because of slight buoyancy for their ascent. For this reason, they descend carrying ballast, and exchange it for slighlty less of the nodules. The ballast, therefore, constitutes the energy for nodule ascent.* **14**, *batteries supplying the power for manoeuvring and dredging. Photo courtesy of GEMONOD**

**This work is now being carried out by GEMONOD, a joint venture between the CEA and IFREMER*

variations. The decoupling of these functions by autonomous systems provides complete flexibility, with the use of sensors and self-adaptive control systems.

Teleoperation is also involved in the maintenance of these devices in emergency conditions, and leads to an approach comparable to teletrans (see Chapter 4.1), in that remote dextrous work, transport and transfer are all implied in the optimized general processes.

Conclusions

This overview of underwater teleoperation is intended to illustrate the areas in common between two of the leading areas of teleoperation application, namely nuclear teleoperation and underwater teleoperation. The similarity in the system analysis is more striking than the differences based on the operation, risks and habits of users in both fields of application, and helps prove the common basis of teleoperation, which is becoming more apparent with the developments in each field.

Instead of developing specialized machines with increasing versatility and rejecting the potential of far more polyvalent systems, the pragmatic procedure should incorporate matrix analysis allowing reasonably polyvalent classes of equipment in each technology. This would provide an overall view of the research and development programmes that are lacking, which, unless the delay in their deployment is corrected, will lead to technological inadequacies and weakness in specific areas.

Among the essential points for development in teleoperation are the problems of power storage (common to manned underwater vehicles and teleoperators), supply and signal transmission cables (common to diving bells and ROV), dextrous manipulators and television (common to both undersea and nuclear) systems for multiplexed encoding of signals and finally reasonable but realistic introduction of computers and sensors in all these systems.

The specific technological problem common to all underwater devices remains, of course, the crossing of the air-water interface and lifting with compensation for the effects of the up and down movement due to waves.

Chapter 4.3

Space applications of teleoperation

As considered for nuclear and underwater teleoperation, robotics applications in space is also a subject for study, both from the point of view of remotely manned systems and the presence of man in space. The large distances involved, however, in outer space teleoperation, even in low orbit, can give rise to even more communication problems than encountered in underwater work. In particular, there are transmission delays caused by the time taken for the signals to cover the distance and for the data to be processed (whether these relate to sensor signals or control signals). The field of outer space exploration and now exploitation has led to developments in the robotic process in the same way that nuclear research and industry earlier put pressure on the development of the technology for remote systems.

The various functions involved in outer space applications are summarized in terms of the demands they can make on teleoperation. As manipulation plays a major visible role in teleoperation, the teleoperator system which seems to be truly operational to date is the manipulator on the American space shuttle, but satellites and long range planetary probes must be said to rely on teleoperation which is mainly concerned with vehicle systems.

4.3.1 Planetary exploration and experimentation

The remotely manned space probes mentioned above required, and in fact brought about, a considerable contribution to advanced robotics as well as advanced teleoperation. Therefore, they will not be discussed further.

The planetary missions which have landed safely on the Moon (Surveyor and Lunakod), Mars (Viking) and Venus (Venera) were first and foremost concerned with transport and transfer systems, limited to the function of soft landing and observation. Figures 1.1.4.(1) and 1.1.4.(3) showed the only two systems equipped with manipulators. Lunakod [see Figure 1.1.4.(2)] is the first remotely manned vehicle to have travelled around on another planet with the Moon rover [see Figure 1.6.7.(6)], and vehicle concepts being developed for exploration

1
4
2
3
a

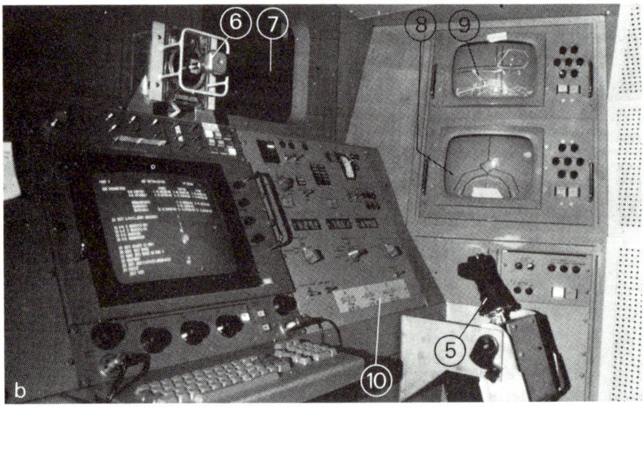
6
7
8
9
5
10
b

Figure 4.3.2.(1) *The Remote Manipulator System (RMS) for the American space shuttle (SPAR, Canada, 1980)*
(a) *RMS retrieving a satellite:* **1**, *15 m long articulated arm;* **2**, *end effector ensuring that prehension is achieved by cable tightening on* **3**, *docking part on the satellite being handled;* **4**, *camera mounted on the end effector*
(b) *Control station on board the space shuttle:* **5**, *lever controlling movements according to the direct view frame (with reference to the cargo bay) through the porthole 7.* **6**, *control lever for tip orientation. These levers are identical to those that control space capsules.* **7**, *porthole for viewing the cargo bay;* **8**, *television monitor providing the equivalent of a direct view of the bay;* **9**, *monitor for camera 4;* **10**, *control panel for mode selection of velocity control according to the frame of view (cargo bay and tip) or relative to an approximate velocity of the satellite to be retrieved. Photo courtesy of SPAR Aerospace, Canada*

on Mars in the future and its satellites were shown in Figures 1.6.7.(1) and 1.6.7.(11).

Apart from the two Moon teleoperators equipped with signal transmission between earth and the Moon of a few seconds, which allows direct remote control, these systems fall into the category of advanced teleoperation and/or advanced robotics (mobile robots) because they must be capable of a high level of decision making autonomy in the choice of their route between fixed halts.

The progress towards autonomy in the choice of route for a mobile robot may be more important than the study of legged locomotion, but it would be true to say that the two aspects complement each other. The mobile robot is as much a symbol of the autonomous control system generating strategies as the articulated arm was earlier a symbol of robotics, by presenting a control problem based on complex processes. Active Suspension Vehicles (ASV), with or without wheels, demand a solution for high-level control with coordination of the highly redundant legs, and particularly for *local self-adaptability* to the terrain. The mobile robot must *know* and *understand* the terrain as far as the horizon (and may extend to its map).

The demands of these vehicles have already been mentioned in Section 1.6.7 in terms of mechanisms and optimal structuring; in parallel with this, there is a great demand for autonomous control systems based on an optimal sensor structure, and because of this the control of an end effector with proximity sensors, as seen in Figure 1.7.11.(2) and mentioned in Section 2.2.2, can be applied in roughly the same way for any mobile system on land, in water or in space. This subject enters the field of advanced teleoperation, in the area of intersection with advanced robotics, also known as third generation robotics [see Figures 1.2.6.(3) and 1.2.6.(4)].

4.3.2 Satellite maintenance and servicing

The development and success of the American space shuttle made the public aware of the remarkable work carried out by SPAR in Canada, which resulted in the production of the Remote Manipulator System (RMS), which, at 15 m long, is the largest in the world (Taylor, 1978). This remote manipulator uses unilateral rate control coordinated by computer, while taking into account all the conventions of aeronautical and space technology, particularly in relation to the control joysticks used by the operator, which are identical to those used for the manual controls in spacecrafts. Figure 4.3.2.(1) shows this manipulator in action to capture large satellites. For maintenance work that requires the replacement of modules inside a large satellite, the RMS has the additional role of acting as a large

Figure 4.3.2.(2) *Solaris satellite for materials production (CNES project, France, 1982):* 1, *shuttle vehicle providing raw materials ending its approach, and held by the manipulator* 2 *for final docking;* 3, *furnace for materials preparation;* 4, *shuttle containing finished materials, ready to return to earth;* 5, *second manipulator used to maintain the furnace;* 6, *solar panels providing the power to run the furnace;* 7, *antenna pointing towards a geostationary relay satellite. Photo courtesy of CNES (Centre National d'Etudes Spatiales)*

transporter transporting a small platform, on which the astronaut stands, to a position close to the task. The astronaut, not the manipulator, is responsible for the dextrous execution of the task itself. This illustrates one possible division of tasks between the manipulator and man in extra-vehicular activity (EVA). On the other hand, a dextrous slave subsystem equipped with two or three arms is envisaged for the RMS.

Figure 4.3.2.(2) shows the French Solaris project for materials production out of gravity at the moment the space shuttle leaves the station to transport the produced material to earth. A number of robot manipulators will ensure that the various parts of the station run smoothly: final rendezvous, introduction and removal of materials and local maintenance. This project could form a part of the European space programme before the end of the century.

A project concerning a telecommunications satellite maintenance manipulator, controlled from a great distance, has also been undertaken in the USA (the Ross Servicer), and there are plans for an experimental European craft to be launched into orbit in the 1990s. In connection with this topic, it seems appropriate to emphasize the work of two American companies, Grumman and Martin Marietta, and the European company Matra (France).

4.3.3 Assembly and maintenance of large space stations

This line of study is currently being developed in the USA as a prelude to the future construction of large structures in outer space. To cooperate with mechanized deployment machines, the idea of the operator is retained, and in the systems envisaged the operators would be housed in a pressurized cabin, which allowed good local visibility, either free or carried at the tip of a large transporter [see Figure 4.3.3.(1)], known as the Cherry Picker. The name is derived from its similarity to the platforms on large articulated boom transporters used for picking fruit and other types of work. The same basic concepts apply, but in this example the operator manipulates arms of varied levels of dexterity from inside the cabin viewing the scene directly, thus performing the task in a similar way to the tasks operated from a submarine equipped with manipulators.

The American space station (SS) that may be built around 1992 at a cost of US $9,000 million will house 6 to 18 persons. It will be 120 m long and made up of 5 to 10 modules. Co-orbiting automatic platform(s) and free-flying servicing vehicle(s) (manned or remotely manned) will be associated with the SS. The European Space Agency (ESA) intends to cooperate with this American project and the Columbus project, first by attaching some modules to the SS and then a smaller free manned station is envisaged.

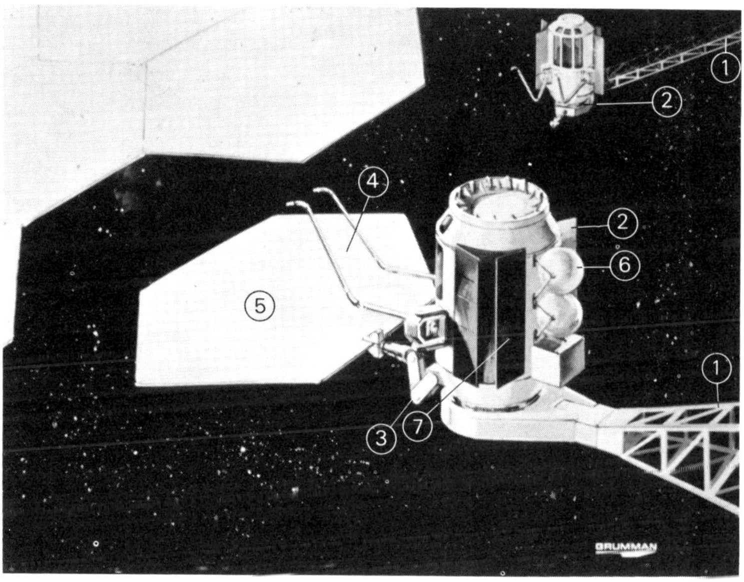

Figure 4.3.3.(1) *Cherry picker concept for the assembly of space stations (Grumman, USA, 1982):* 1, *beam of large transporter integral with framework of the station;* 2, *pressurized cabin similar to those used on large cherry pickers for laying pylons or trimming trees, hence the name of this device;* 3, *manipulator (left arm);* 4, *dextrous bilateral servo manipulator;* 5, *part of a vast mirror, ready for assembly (solar power-station);* 6, *operator oxygen cylinders;* 7, *radiative panels to avoid overheating. Photo courtesy of Grumman*

In these projects, manipulators will be developed for assembly, docking and servicing.

Conclusions

The application of teleoperation in outer space, in common with the previous examples, provokes considerable discussion over the real needs and uses of having the operator present at the site of the task. One important factor in favour of this requirement is the signal transmission delay, which involves the use of highly advanced teleoperation techniques for remotely controlled systems. For this reason, work in low orbit, which is the closest possible approximation to any practical application, is being developed using either an operator in a space suit for EVA or in a cabin for close teleoperation in the medium- and long-term American and European projects, while manipulators used for servicing telecommunications satellites and materials production stations are the most likely application from now until the end of this century for long distance advanced teleoperation. Among the tasks entrusted to the manipulator, the capture of satellites and/or their repair is currently seen as the likeliest area for development. The European programme is leading to the launch of manned spacecraft, which is now possible with the Ariane booster. In parallel, European efforts are devoted to robotics and teleoperation technology in space and will use the best available knowledge on advanced teleoperation and advanced robotics.

Chapter 4.4

Medical applications of teleoperation

This subject merits an entire book to itself in view of the potential applications. Jack Guittet (1985I, II) has published an account of the equipment that can help handicapped people. The application of robotics to medicine in general could relate to the operating of hospitals as much from the point of view of personal assistance (handling), surveillance and care as from that of rehabilitation in which robots with real physical action play an important role. Robotic techniques and/or the judicious use of computers and sensors are already of, and will be of increasing, assistance to patients with sensory handicaps and to doctors (computer-aided diagnosis).

Robotics and teleoperation can also help handicapped people in real life, sometimes allowing them to be reinstated into society. Only the strictest applications of teleoperation are considered here, and these will make considerable use of CAT. Into this category fall both applied teleoperation to help hospital staff handle patients with motor handi caps, and manipulation techniques for use in surgery (micro-surgery, in particular).

4.4.1 Teleoperation for handling and transporting patients in hospital

The pioneering work in this field was conducted by MITI in Japan at the Mechanical Engineering Laborabory (MEL) where the feasibility of a *patient handling robot* has been demonstrated, in addition to a multidirectional platform with the potential use as a wheelchair for the disabled (Nakano and Koyachi, 1983).

4.4.2 Applied teleoperation for patients with motor handicaps

4.4.2.1 ASSISTED MANIPULATION

Hand and arm prostheses

Hand and arm prostheses are remote manipulators in themselves. They make use of the resources available to supply control while respecting the structural form of the missing limb. This is the case, for example, with artificial hands using electromyographic control by the remaining muscles which would have controlled hand movements. These muscles can directly control an approximately equivalent movement of a prosthetic hand. This particular example, in which the manipulator is in physical contact with the operator, shows similarities to the mechanical master-slave manipulator, but the absence of certain of the arm actuations makes this contact with the manipulator unilateral. Nevertheless, touch (sensitivity of contact) between the prosthesis and the partially amputated arm allows an indirect kinaesthetic feedback, which is essential for control.

Motorized ortheses

Motorized ortheses are used in cases where the limb is present, but paralysed, to train its movements. As a result of the important work conducted at the Case Institute of Technology (USA), the Rancho Hospital group in California has developed a motorized arm orthesis with very high-level control, and has introduced the use of robotics to help people with motor handicaps. This small series of motorized ortheses was produced, and some were used by certain Californian groups as experimental manipulators, before the introduction of better adapted motorized manipulators.

Telethesis

Telethesis is another form of remote manipulation, which was introduced more recently (Guittet *et al.*, 1978) to deal with cases of severe paralysis; the system by which a real remote manipulator is controlled directly by the handicapped person has come to be known, by association with the techniques mentioned above, as *telethesis*. To the best of our knowledge, five teams in succession have become interested in this approach: Rösler's group in Heidelberg (Rösler and Paslack, 1976; Paslack and Rösler, 1973, 1975) [see Figure 1.4.1.(9)]; a group at John's Hopkins University (USA) (Seamone *et al.*, 1978); Jack Guittet's group as part of the Spartacus project; a group at the JPL (USA) mounted an arm developed by the Veterans Administration on a wheelchair with voice control; and Leifer and Roth at Stanford University (USA) (Leifer, 1981; Leifer and Sachs, 1979; Leifer *et al.*, 1978).

It is important, at this stage, to restate the construction complexity of motorized ortheses, in which there is no spare space, and the motorization must be able to transport the dead weight of the arm. In the case of the motorized prosthesis, the available volume and weight must be less than, or equal to, that of the missing limb. In addition, an important aesthetic criterion must be fulfilled. In both cases, it is also of the greatest importance that the mechanisms should be very perfectly adapted to the needs of each patient. Telethesis presents a different problem in that part of the control system must be adapted to each user. The mechanism is a standard remote manipulator system, with the difference being that the operator's control capacity is limited by his handicap. It is possible to use analytical control, with a long duration of execution, as in the system developed by Rösler (Rösler and Paslack, 1976), with levers controlled by the tongue.

The analytical control employed in the Johns Hopkins system was so slow that a selection of programs was finally used. Hamonet *et al.* (1978) in France concentrated on the use of electromyographic control.

By using high-level robot programming languages, the Stanford project concentrated more specifically on the use of voice control, which allows the possibility of communication without using gestures. Nevertheless, the non-analog character of the language used leads to coding the table surface on the bases of X Y coordinates to identify squares. The lack of intrinsic safety means that the commands received must be validated by speech synthesis, and this increases the execution time for each command. This is why the group has recently been looking into the transparent kinaesthetic control, which has been adopted since the start of the Spartacus project. The Stanford group has based its experiments on the Puma.

The other process remains within the tradition of teleoperation, which has formed the subject of this volume. This is the approach adopted in the Spartacus project, which makes full use of the instinctive potential for control (the concept of transparency) and concentrates on the remaining head movements (generally two or three altogether). Other remaining arm motions and the voice allow a certain number of modes to be controlled, such as coordinated control of the six DOF of the manipulator and its gripper and control of wheel movements [see Figures 1.4.5.(9) and 1.5.0.(3)].

This fascinating project has highlighted the main concepts of teleoperation using computers in parallel, which characterize our methods, and although they approach supervisory control described by Sheridan, they remain distinct. This case illustrates the concept of cooperation robotics (André, 1980), in an extreme way. The MAT.1 manipulator (Vertut, 1978), which was specially developed for these experiments, is currently being very successfully evaluated clinically (Pannier, 1981; Kwee, 1983; Kwee and Pannier, 1981), using, in addition, control for additional commands. A subject was completely able to cook and eat a meal. Figure 4.4.2.(1) shows a child celebrating his reeducator's 'birthday'.

Environmental control

This is based on an approach which makes use of special machines which can, once again, be considered in opposition to the multipurpose

Figure 4.4.2.(1) *A patient celebrating his reeducator's 'birthday'*

machine approach. Achieving a reasonable level of versatility leads to the juxtaposition of a large number of systems, such as page turners, adapted telephones, typewriters and lighting control, all in analytical mode with successful voice control. This has the unique advantage of representing a progressive investment as opposed to the overall investment of a telethesis system, but without providing a capacity for manipulation and transport.

4.4.2.2 ASSISTED WALKING

Following on from the work of Tomovic the Yugoslav, Rabischong applied the problem of assistance to those with paralysis of the lower limbs using a motorized orthesis. His original idea (Rabischong *et al.*, 1978; Hill, 1976I) consisted of controlling the orthesis by unilateral positional servocontrol using two exoskeleton legs worn by the patient [see Figure 4.4.2.(2)]. The second version, currently being used experimentally, is hydraulically powered and was produced by Renault. This system is highly promising for training limbs; the extension towards autonomy on the basis of a kinematic computer model of walking is envisaged in the long term. The patient would use two walking sticks.

4.4.3 Telesurgery

This is an attractive example of a high-level dexterity application of bilateral remote manipulation with a size scale reduction to between one-tenth and one-hundredth (micro-manipulators used in cell research work at a scale of less than one-thousandth with analytical control).

Jourdain's system for telesurgery has existed for some time in mechanical form, for simple operations using an endoscope (removal of objects from the digestive tract, or certain gynaecological or urological operations). The use of advanced remote manipulation techniques with bilateral servocontrol would allow these approaches to be developed in a highly effective way, both in terms of precision and dexterity.

A conceptual design has been envisaged for endoscopic pneumoperitoneal telesurgery, which can be used to avoid bacteria infections associated with opening the abdominal cavity in the operating theatre, and to greatly reduce shock (Alexander, 1973). To the best of our knowledge, this work has not resulted in practical developments.

A special manipulator for eye surgery is, however, currently being developed by our colleagues in Lille (Lepers, 1982, 1983I, II). The ARA project intends to couple the arm with a MA23 manipulator.

By adding to these two examples, the fields of ear surgery and neurosurgery, it becomes clear that there is tremendous scope for the development of new techniques in all these and more areas.

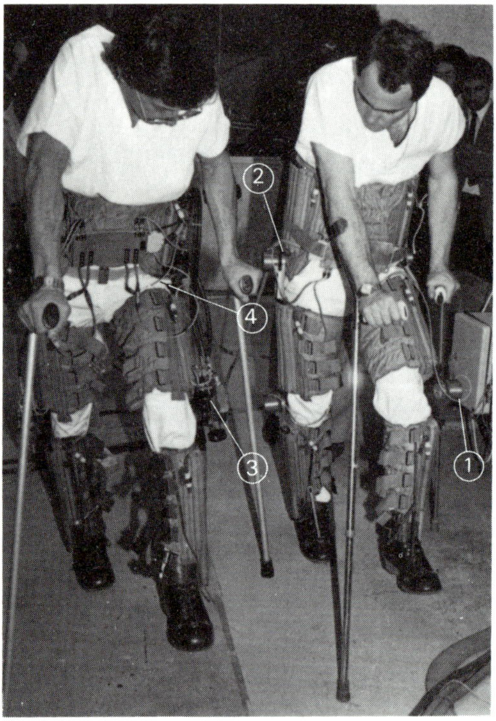

Figure 4.4.2.(2) *Active modular orthesis for lower limbs (OMAMI) (Rabischong, INSERM, France, 1983):* **1 and 2**, *potentiometers for the master orthesis, worn by the patient;* **3 and 4**, *slave hydraulic actuators for the patient. Contention on the segments is ensured by the presence of inflatable pieces reinforced with strips of composite material (carbon fibre). The hydraulic system was produced by Renault, the orthesis by Aerazure. The kinematic walking model, developed by the Automation and Microelectronics Laboratory, Montpellier (LAMM) is intended to be used and to give the patient greater autonomy. Photo courtesy of INSERM*

Conclusions

This field would benefit greatly from the knowledge specific to the fields of standard and advanced teleoperation, and from the scale effect which is favoured by the bilateral servo manipulator system, which is both reduced in size and has amplified force feedback. The use of sterile enclosures (Isotechnie-La Calhène) in association with these systems could also lead to the use of the teletrans concept (Vertut *et al.*, 1984V).

Chapter 4.5

Industrial applications of teleoperation

This chapter concerns the tasks that are unsuitable for the industrial robot and even the next stage models. These tasks are generally not repetitive and difficult to model, and require almost permanent intervention by a trained operator, particularly in forging and metal work.

4.5.1 Applications in metallurgy and forging

This relates, of course, only to free forging operations, carried out under direct viewing conditions (without a die, unlike die forging), to produce cubes, cylinders and other rough shapes. The power hammer is used to pound the block of metal directly, which is usually large. An example of this was shown in Figure 1.3.10.(1), in the form widely used for metals of reasonable size. There are specialized large capacity forge manipulators, allowing a very high quality of work to be carried out, but the Man Mate (Andromat) appeared to be a very promising tool for the free forging of various sizes of metal ranging from 100 kg to 1 tonne [see Figures 4.5.1.(1) and 1.5.0.(3)(d)]. There are more than 100 units in use throughout Europe.

There are other examples of applied teleoperation in the metals industry, including systems for cleaning melting vats, using a Japanese remote manipulator mentioned by Köhler (1981, p. 535), deburring and grinding (Man Mate).

4.5.2 Public works applications

In some cases, machines used for public works fall, more and more, into the category of teleoperation, and the development of sophisticated machines with automatic functions or bilateral control is to be expected in this field. This is the case, for example, with the bulldozer by Komatsu (Japan) in which a laser beam is used to control the height of the blade, thus allowing levelling to be carried out in a single pass, leaving the driver free to concentrate solely on driving the vehicle.

The DRET (the French equivalent of the Defence Advanced Research

Figure 4.5.1.(1) *Free forging using the Man Mate (Andromat) [see Figure 1.5.1.(3)]:* 1, *1 tonne pile hammer;* 2, *metal block being pounded into a cylinder;* 3, *Man Mate (Andromat) gripper controlled by the operator using the movements of the master arm at a scale of 1/10, with force at a scale of 1/100. Gripper orientation is switch controlled, and only position is bilaterally servocontrolled. As a result, the operator can hold the block between the fingers of the gripper, parallel to the axis of the cylinder, and rotate it easily.* 5, *reheating furnace from which the block can be removed directly because of the large coverage of the manipulator's movements. A single operator, having found a more efficient way of working, can achieve the same type of production as a team of three, with a 30 per cent improvement in rate of work. A cabin may protect the operator from the outside conditions (noise, heat and dust). Photo courtesy of Jean Vertut*

Project Agency or DARPA in the USA), before the introduction of the Man Mate, carried out some experiments in France using a bilateral servocontrolled mechanical shovel which was capable of giving improved results even without visibility in a flooded trench.

4.5.3 Mining applications

A British colleague, M. W. Thring, has long been pressing for the use of teleoperation in mining, partly for exploitation, but mainly for the maintenance of automated systems in coal mines. He has consequently suggested that ventilation, which is vital to the presence of men working underground, could be abandoned. This reallocation of investment in mining as a whole may deserve careful consideration. Jean Vertut's laboratory at Saclay has verified the feasibility of a number of maintenance operations, and in particular the replacement of diamond cutting tools using the MA23 manipulator.

Two characteristics of mining environments are dust and the confined space in the corridors where work is carried out, which impose demands on the designer. Figure 4.5.3.(1) shows one of Thring's ideas (1976), and a design produced by NASA for a working machine is given in Alexander (1973).

The American company MBA developed an experimental manned teleoperation vehicle (Foote and Dompe, 1978). To the best of our knowledge, it has not been used in mining applications.

Wireless remote control at viewing distance is already operational for vehicles and loaders in some French mines.

The French RAM project is considering an advanced robotics approach to coal mining.

4.5.4 Handling objects that cannot be modelled by computer

Figure 1.4.1.(8) showed an example of teleoperation applied to the replacement of facing plates in larger grinders.

Among the applications not concerned with mechanical industrial production, the rapid handling of objects with varied (and variable) shapes, such as meat, has been the subject of experiments at the Automation Centre, Lille, using the Man Mate (Andromat).

This type of revolution in the working habits, even if the technology offers many advantages, will usually meet with opposition, despite the fact that operators are required at the controls in areas such as remote manipulation. It was in connection with these specific applications of teleoperation that our colleague Vidal (1983) first studied automating the change over from the manual mode to the automatic mode, that is,

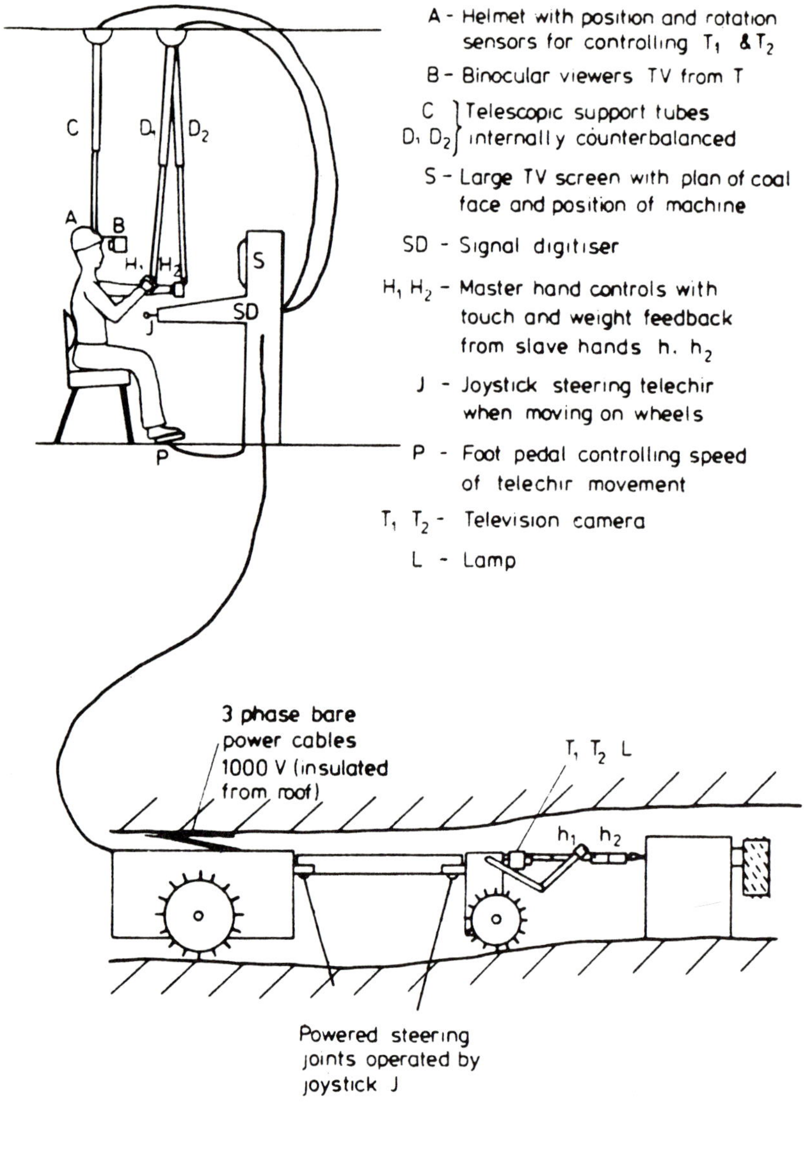

Figure 4.5.3.(1) *M. W. Thring's concept of mining teleoperation (Queen Mary College, London, 1970): From Thring (1983) Robots and Telechirs, p. 267, Ellis Horwood*

the establishment of the compound mode of work which is now being studied at the Advanced Teleoperation section of the ARA project (Jolly, 1982).

4.5.5 Work on high voltage power lines

In forging and grinding, to increase the force and the reach is the main goal, despite the hostile environment. This second property of teleoperation application is based on the hostile environmental aspect added to the specific location of tasks. The DEMS manipulators by General Electric (or the AS30 by AST in West Germany) have been used experimentally in operations concerned with maintenance, from the ground, of high voltage power lines, with successful results. Sumitomo, in Japan, is approaching this application under the auspices of the Jupiter project.

Conclusions: promises of industrial servicing robotics

The variety of industrial tasks that may involve the use of teleoperation does not allow easy classification. Either the hostile environment aspect is foremost or the amplification of force allows high dexterity manual control in cases where automation is too difficult, or both these aspects can be present simultaneously (eg in forging, mining).

In this field a period of introduction of CAT will probably exist before autonomous robots could have the recognition and decision making capacities needed for executing difficult tasks.

Part of this major step is now being made (entering industrial servicing robotics), with new possibilities for the automation of tasks which were previously impossible to automate.

The new fields are industrial cleaning and inspection. The Paris Transportation Authority (RATP) started an ambitious programme to develop cleaning robots. The first phase concerns the cleaning of buses and trains. The next phase envisages a robot for cleaning underground platforms and stations. The work of Midirobot, AID and the CEA must also be mentioned here.

Inspection mobile robots for hazardous chemical facilities are also envisaged. These will be similar to the projected routine inspection robots in nuclear power plants.

Another new field is agriculture. A forest exploitation mobile robot is now under development in France.

These three applications are concerned with advanced vehicles (mobile robots) and may require in the first step a human supervisor for several units. They all form part of the RAM project.

Handling robots to serve automatic mail sorting machines, replacing post office workers, are being used experimentally by La Calhène and Renault in France and therefore also warrant a mention. Robots that will be used to select and load video cassettes (or disks) in large video libraries are also being developed by La Calhène and AID.

Such robots are usually fixed arms or arms that can make limited displacements, in opposition to mobile servicing robots. Again CAT techniques and a human supervisor may be used in the first step.

This new field of non-manufacturing robotics, or servicing robotics, is opening up just when the present work is finishing. In France, the CEA is devoted to this field, as it is similar to remote maintenance in hostile situations (eg nuclear, underwater, space).

Chapter 4.6

Applications in security and civil protection

The aspect of working in a hostile or dangerous situation is prominent in tasks that have until now been carried out by hand, involving the obvious dangers of fighting fires, defusing bombs, major emergencies and disasters and police work.

4.6.1 Fighting fires and saving lives

To the best of our knowledge, Japan was the first country to develop remotely controlled vehicles carrying fire hoses which could be moved and aimed using a direct view of the scene, and to use them under experimental conditions. This application deserves to be more generally developed as much for the protection of firemen as the improved efficiency of the system, since it would be possible for this type of vehicle to approach the scene of an accident under conditions that would prevent any human presence. It is easy to imagine the contribution that could be made by robots able to enter buildings during emergencies when the life of rescue workers would otherwise have to be risked. This sort of situation would, however, demand high levels of performance in the speed and precision of decision making, which are not currently feasible with the present state of remote systems technology.

Work continues at the JIRA in Japan on the analysis of safety problems, and three systems have been devised for exploration, guiding people to safe places and shielded transport (Kobayashi and Nakamura, 1983). This theme is part of the Japanese Jupiter project.

Another device likely to contribute to the protection of people in emergencies is a wall climber that can move across the facades of buildings, for example, controlled directly or remotely, which could be derived from the vehicle shown in Figure 1.6.7.(4). This type of device could be used, not only for saving lives in emergencies, but also for repairs to offices or residential buildings.

4.6.2 Applications in bomb defusing and disposal, and the police force

To the best of our knowledge, it was Edwin Johnsen who first started to develop this application, which has resulted, in recent years, in the

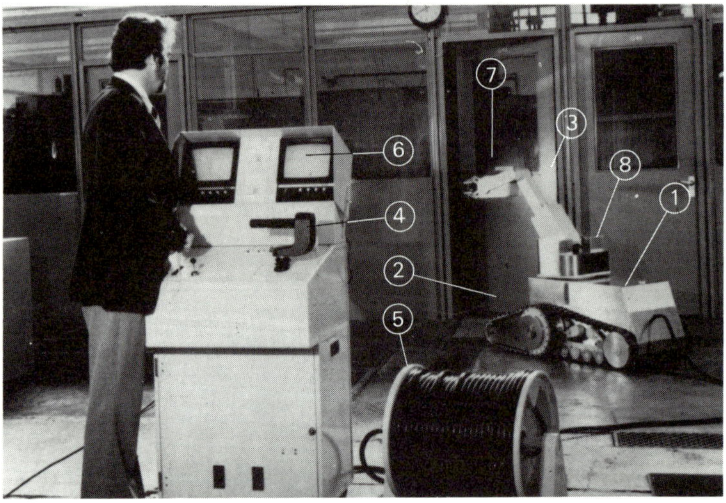

Figure 4.6.2.(1) *Spider teleoperator (AERE,* Harwell, Great Britain, 1978):* **1**, *caterpillar-tracked (0.92 × 0.70 m) vehicle;* **2**, *orientable additional caterpillar tracks allowing stairs to be negotiated (length 0.3 m);* **3**, *motorized manipulator: length 1.2 m, capacity 15 kg, with eight independent movements under constant velocity control using the handle* **4**; **5**, *spool containing 100 m of multiconductor cable;* **6**, *television monitors;* **7**, *camera on the gripper block;* **8**, *cameras on the manipulator turret. This system is intended for nuclear intervention, and it is also suitable for police applications or for defusing explosives inside buildings. The two redundant movements are a telescopic extension of the gripper and the extension of an upper arm, which is already equipped with a rotation about its axis, which is impossible to place on the small fork-shaped forearm. Photo courtesy of the United Kingdom Atomic Energy Authority (UKAEA)*

* Atomic Energy Research Establishment

use of a number of systems in the USA for handling and disposing of outdated or unsafe munitions. On the other hand, the defusing of explosives in anti-terrorist activity presents a different set of problems, among which is that of exposing bomb disposal experts, who cannot always predict the methods used for constructing home-made bombs, to extreme danger. Occurrences in Northern Ireland have led to the development of a number of simple remote vehicle-mounted manipulators, which generally have a short life span, but can save many human lives. These devices are mainly used to approach booby-trapped cars or suspicious looking objects, thought to be explosive.

Figure 4.6.2.(1) shows a British device, developed by a nuclear research group and, interestingly, capable of climbing stairs. Also worthy of note are the Hobo device, used in France, the MF3 developed by Köhler in West Germany [see Figure 1.6.7.(10); Köhler, 1981, pp 570 to 612], the Gripy system by La Calhène and a device in the process of being developed by AID.

Considering the high risk of these systems being destroyed, there is a natural tendency to reduce the price, although this is generally the cause of a major reduction in efficiency. This is another phenomenon which does not favour technological progress, but indicates above all a difference between devices for quasi-military use and those for intervention in case of accidents or natural disasters. Vehicles that would be needed for police applications, where buildings must be entered remotely, would also fulfil the requirements for fighting fires and saving lives, mentioned above. These problems will only be satisfactorily solved in sophisticated machines, since now it is more difficult to perfect the means to achieve mobility and agility than dexterity.

4.6.3 Applications of legged locomotion

This section is intended as a summary of the reasons for the possible development of legged locomotion, bearing in mind that the discontinuous character of the contact with the ground is very favourable to displacement over terrain that is uneven and cluttered with obstacles. Out of all known terrain on earth, 25 per cent is inaccessible to conventional offroad vehicles. Legged locomotion should make a good proportion of these areas accessible, and it is, therefore, of interest not only to the military, but to forestry workers, and for public works etc. Another justification for this type of vehicle concerns locomotion over areas which are frozen to a certain depth all year round, and on which any continuous vehicle contact can lead to serious erosion after a number of years. This accounts for the efforts made by Soviet groups in connection with the vast Siberian plains [see the model shown in Figure 1.7.9.1(10)].

When a vehicle of this type can transport its own pilot (using CAT techniques), there will be a high level of cooperation between the sophisticated automated system and the reflexes and decision making ability of the pilot, working in real time with a partial work load (and in part time). Vehicles of this type, controlled remotely, should be the subject of development and will be of great significance to all areas of teleoperation, particularly for intervention in nuclear, underwater and outer space facilities, or for fighting fires.

Chapter 4.7

Conclusion

This discussion of teleoperators and robotics since its origins represents 40 years of a tradition in which the relationship between mechanical- and energy-based techniques and the operator has gradually developed. In the last 15 years, the slow introduction of computers has demonstrated the far reaching possibilities of *cooperation robotics*, which basically has the same aims as *substitution robotics*, but maintains contact with man, respecting his place in the system, and the importance of his actions in the execution of difficult, critical or important tasks. This is why advanced teleoperation takes its place more in the field of non-repetitive tasks than in the sphere of tasks that require specialized or flexible automation to provide a means of repetitive performance.

It so happens that this emphasis occurs in *hostile environment robotics* when the hostile environment is entered during the non-repetitive period of operation for maintenance, repair or dismantling work, or when accidental occurrences in facilities render the environment hostile.

This double aspect does not perhaps help in justifying the development of teleoperation for these applications, but the fact is that hostile environment and cooperation robotics are very costly areas of research when considered in terms of how infrequently (if ever) their use is essential. This has caused these areas to be neglected when they should ideally be taken into consideration, for example, when designing new facilities.

A third approach, which reflects the ambivalence felt with respect to the economics of robotics and teleoperation research, is that of *service robotics*. The service robot can often pay for itself after being used only once, but it constitutes a form of insurance which is costly only until there is an accident. This new type of robotics can raise a new economic approach because of its non directly producing nature.

It has been one of our main intentions, in this series, to stress the particularly broad approach of teleoperation in robotics, allowing problems to be approached in a rather more general way than is usual in the current context of robotics. This has led to considerable efforts to reduce the resistance to innovation, which has so far opposed the use of applied teleoperation in industry to the extent it merits, for more difficult tasks demanding a high degree of flexibility. This is why so much emphasis has been laid on the importance of common ground in widely different applications, to reinforce the idea of a common root in research and development, encouraging full use to be made of such a promising technique in a variety of applications, when the effort required could not truly be justified in relation to any one singly.

It is important to state that computer-aided or advanced teleoperation, resulting in the same level of ultimate autonomy as advanced robotics, makes a considerable contribution to the development of advanced robotics because of its emphasis on the human factors, assuming the necessity for a human operator to recover the system, despite his shortcomings.

Advanced teleoperation, hostile environment robotics and service robotics can, therefore, be presented as disciplines that deserve investment for research and development, even more than industrial robotics, since they can be seen as bringing a major contribution to the development of third generation robotics. This discipline will, no doubt, benefit in years to come from the work carried out on advanced teleoperation which it does not raise directly in its progress towards the more direct removal of man from flexible automation systems.

Bibliography

Acquaviva-Lescoat, D. Thesis, University of Orsay, June, 1983.

Acquaviva-Lescoat, D.; Fiori, N. Une tache de télémanipulation est-elle décomposable en éléments indépendants? Soumis au Comité de lecture du Travail-Humain.

Advanced Fuel Recycle Program 1978 *Experimental Facility Interim Design* Report ORNL AFRP 78/6, Oak Ridge National Laboratory, October, 1978.

AFNOR Z 68 101, 1983.

Albe-Fessard, D.; Tyc-Dumont, S. Fonctions somato-sensibles. *Physiologic* Volume 2, 3rd Edition, C. Kayser, Ed., Flammarion, Paris, 1976.

Alexander, A.D. III Impact of telemations on modern society. *Proceedings of 1st RO.MAN.SY* Udine, 1973, p. 121.

Alexander, H.C. Remote systems applications at Three Mile Island Unit 2. *Proceedings of Robotics and Remote Handling in Hostile Environments, National Topical Meeting ANS* Gatlinburg, 1984, p. 169.

Amos, B.; Wang, M. Stereo television viewing for remote handling in hostile environment. *26th RSTD* 1978, p. 358.

Ananiev, A.N.; Nakhapetjan, E.G. Experimental tests of industrial robots machinism. *Proceedings of 4th RO.MAN.SY* Udine, 1981, p. 416.

André, G. Thesis, October, 1983, IRISA, Rennes, June, 1984.

André, P. *Cooperation Robots* Thesis, University of Besancon, March, 1980.

Arai, T. *et al.* A stair-climbing robot for maintenance: A MOOTY. *Seminar on Remote Handling Equipment for Nuclear Fuel Cycle Facilities, IAEA* No. SR103, IAEA-OECD Nuclear Energy Agency, Oxford, 1984.

Auchapt, P.; Streiff, G.; Vertut, J. Association of remote dexterity and remote lifting for maintenance in fuel reprocessing industry. *Proceedings of Robotics and Remote Handling in Hostile Environments, National Topical Meeting ANS* Gatlinburg, 1984, p. 197.

Baldwin, W.C.; Miller, I.V. *Multilegged Walker* Final Report, Space General Corporation, El Monte, California, January, 1966.

Barabaschi, S. *et al.* An electrically controlled servo manipulator. *Proceedings of 9th HLE* 1961, p. 143.

Barbier, R. *et al.* The Spartacus Laboratory. *Bulletin INRIA* No. 74, December, 1981, p. 17.

Bartholet, T.G. The first 'functionoid' developed by Odetics Inc. *Proceedings of 1983 ICAR* Tokyo, 1983, p. 293.

Bartholet, T.G. ODEX I. A new class of mobile robots. *Proceedings of Robotics and Remote Handling in Hostile Environments, National Topical Meeting ANS* Gatlinburg, 1984, p. 261.

Bejczy, A.K. Distribution of control decision in remote manipulation. *Proceedings of IEEE Conference on Decision and Control* Houston, December, 1975.

Bejczy, A.K. Effect of hand sensors on manipulator control performance. *2nd RMS Conference, MMT* Volume 12, No. 5, 1977, p. 547.

Bejczy, A.K. Manipulation of large objects. *Proceedings of 3rd RO.MAN.SY* Udine, 1978, p. 301.

Bejczy, A.K. Sensor and computer-aided control of manipulators in space. *MIDCON 1979* Chicago, November, 1979.

Bejczy, A.K. Remote applications of robots. *Proceedings of 1983 ICAR* Suppl. Tokyo, September, 1983, p. 1.

Bejczy, A.K.; Corker, K. Manual control communication in space teleoperation. *Proceedings of 5th RO.MAN.SY* Udine, 1984, Kogan Page, London, 1985.

Bejczy, A.K.; Dotson, R.S. A force torque sensing and display system for large robots. *Proceedings of IEEE Conference* No. F1, April, 1982.

Bejczy, A.K.; Handlykken, M. Generalization of bilateral force reflecting control of manipulator. *Proceedings of 4th RO.MAN.SY* Warsaw, 1981I, p. 242.

Bejczy, A.K.; Handlykken, M. Experimental results with six degrees of freedom force reflecting hand controller. *Proceedings of 17th Conference on Manual Control* UCLA, June, 1981II, p. 465.

Bejczy, A.K.; Salisbury, J.K. Kinesthetic coupling between operator and remote manipulator. *Proceedings of International Computer Technology Conference, ASME* Volume 1, 1980, 1, p. 197.

Bejczy, A.K.; Zawalki, R.L. Computer-aided manipulator control. *Proceedings of 1st International Symposium on Mini- and Micro-computers in Control* San Diego, 1979.

Bejczy, A.K. *et al.* Force torque control experiments with the simulated space shuttle manipulator in manual control mode. *Proceedings of 18th Conference on Manual Control* Dayton, June, 1982I.

Bejczy, A.K. *et al.* Manual control of manipulator forces and torques using graphic display. *Proceedings of IEEE Conference on Systems, Man and Cybernetics* October, 1982II.

Bekker, M.G. *Introduction to Terrain Systems* University of Michigan Press, Ann Arbor, 1969.

Berry, H.; Rice, O.F. A three-dimensional television system for use in remote servicing. *Proceedings of 26th RSTD Conference* 1978, p. 299.

Bertrand, J.C. Study of operation and testing of robots and remote manipulators. *Proceedings of 2nd ARA Journal* Toulouse September, 1983, p. C1.

Bertrand, J.C. *et al.* Definition of test methods for robots and teleoperators. *Proceedings of 1st ARA Journal* Poitiers, September, 1982, p. 43.

Bessonov, A.P.; Umnov, N.V. The analysis of gaits in six-legged vehicles according to their static stability. *Proceedings of 1st RO.MAN.SY* Udine, Volume 1, 1973, p. 1.

Bessonov, A.P.; Umnov, N.V. Choice of geometric parameters of walking machines. *Proceedings of 2nd RO.MAN.SY* Warsaw, 1976, p. 62.

Bessonov, A.P.; Umnov, N.V. Features of kinematics of turn of walking vehicles. *Proceedings of 3rd RO.MAN.SY* Udine, 1978, p. 83.

Bessonov, A.P.; Umnov, N.V. Stabilization of position of the body of walking machines. *Proceedings of 4th RO.MAN.SY* Warsaw, 1981, p. 447.

Bessonov, A.P. *et al.* Experimental research and diagnosis of robots. Nauka, Moscow, 1981.

Bianchi, G.; Rovetta, A. On the grasping process for objects of irregular shape. *Proceedings of 3rd RO.MAN.SY* Udine, 1978, p. 67.

Book, W.J.; Hannema, D.H. Master-slave manipulator performance of various dynamic characteristics and positioning task parameters. *IEEE Transactions on Systems, Man and Cybernetics* Volume SMC 10, No. 11, 1980, p. 764.

Bouisset, S. Postures et mouvements. *Précis de Physiologie du Travail* 2nd Edition, J. Scherrer *et al.* Eds, Masson, Paris, 1981, p. 29.

Bouisset, S.; Monod, H. Un essai de détermination de caractéristiques anthropométriques en vue de l'aménagement de postes de travail. Etude de 110 'cadres' de la région parisienne. *Trav. Hum.* 1961, **24**, p. 35.

Bouisset, S.; Rebiffe, R. L'aménagement de l'espace de travail. *Medicine du Travail* 2nd Edition, H. Desoille, J. Scherrer and R. Truhaut, Eds, Masson, Paris, 1978, p. 168.

Bouisset, S.; Pineau, H.; Vassal, P. Recherches anthropométriques sur le jeune adulte français. *Biotypol.* 1959, **20**, p. 176.

Brenguier, P.; Bouisset, S. *Proceedings of Sixth Congress of the Biomechanical Society* Brussels, 1981.

Bureau National de Métrologie (BNM) Determination of Robot Performance Paris, December, 1981.

Burgerjon, J.J. *et al.* A solution for remote handling in accelerator installations. *IEEE Transactions* No. NS 24.3, 1977, p. 1580.

Busby *Undersea Vehicle Directory* Busby Associate, Arlington, Virginia, March, 1981.

Campbell, I.C.; Kosyakow, V.N.; Vertut, J. Safety aspects in the design and equipment of hot laboratories. *Safety Series* No. 30, IAEA, Vienna, 1969, p. 25.

Catton, W.T. Mechanoreceptor function. *Physiological Review* 1970, **50**.

Charles, J. *Parc-Eric System for Remote Penetration and Deep Submergence* 1979.

Charles, J.; Vertut, J. La telesymbiotique. *ACOBOR* 1974, **6**.

Charles, J.; Vertut, J. Cable-controlled deep submergence teleoperator system. *Proceedings of 2nd RMS Conference, 1975, MMT* Volume 12, No. 5, 1977, p. 481.

Charles, J.; Marchal, P.; Vertut, J. *Automatisme et Robotique en Recherche Oceanographique. Reunion de l'Union des Oceanographes de France* November, 1981.

Chatten, J.B. Foveal HAT. A head-aimed television system with foveal/peripheral image format. *Proceedings of Symposium on Visually Coupled Systems: Developments and Applications* Aerospace Medical Division, Brooks Airforce Base, Texas, 1972.

Chazalon, M. *et al.* Remote maintenance issues in NET. *Seminar on Remote Handling Equipment for Nuclear Fuel Cycle Facilities, IAEA* No. SR103, IAEA-OECD Nuclear Energy Agency, Oxford, 1984.

Clarke, M. Elements of advanced integrated operator control station. *Proceedings of Robotics and Remote Handling in Hostile Environments, National Topical Meeting ANS* Gatlinburg, 1984, p. 425.

Clarke, M.; Garin, J. Identification of cognitive factors related to remote work performance using closed circuit television displays. *Proceedings of Conference on Manual Control* UCLA, 1981, p. 145.

Clarke, M.; Draper, J.V.; Hamel, W.R. Human factors in remote control engineering development activities. *Proceedings of 31st RSTD Conference* June, 1983.

Clement, G. *et al.* Piade. *CR 21 Reunion des GT Laboratoires Chauds et Telemanipulateurs* CEE MOL, Belgium, June, 1982.

Clement, G. *et al.* Teleoperation and transfer techniques for static manipulation. *Seminar on Remote Handling Equipment for Nuclear Fuel Cycle Facilities, IAEA* No. SR103, IAEA-OECD Nuclear Energy Agency, Oxford, 1984.

Clot, C. Vehicule multibras se deplacant en s'accrochant aux tuyauteries et autres structures. *Personal Communication* 1979.

Coiffet, P. *Modelling and Control* Robot Technology, Volume 1, Kogan Page, London, 1983.

Coiffet, P. *Interaction with the Environment: Robot Sensors and Sensing* Robot Technology, Volume 2, Kogan Page, London, 1983.

Constant, J.A.; Hill, K.J. Roman, a remotely operated mobile manipulator. *Proceedings of 3rd Conference on Industrial Robot Technology* University of Nottinham, March, 1976.

Conter, J.; Valobra, M. Thesis, University of Paul Sabatier, 1980.

Corker, K.; Mishkin, A.H.; Lyman, J. Research issues in implementing remote presence in teleoperator control. *Proceedings of 17th Conference on Manual Control* UCLA, 1981, p. 109.

Corliss, W.R.; Johnsen, E.G. Teleoperator control. *NASA Report* No. SP5070, 1968.

Cowan, S.J. *et al.* Fiber optic communication links for unmanned inspection submersibles. *NOSC* TR652, October, 1980.

Da Costa, L.; Devresse, M. General design of a remote-control multipurpose vehicle for nuclear applications. *Seminar on Remote Handling Equipment for Nuclear Fuel Cycle Facilities, IAEA* No. SR103, IAEA-OECD Nuclear Energy Agency, Oxford, 1984.

Danel, F. Vehicules adherant sur paroi par coussin fluide en depression. *Personal Communication* 1978.

Datta, I.; Kuo, C. The role of manipulators for underwater activities offshore. *Proceedings of 3rd RO.MAN.SY* Udine, 1978, p. 488.

Davidson, H.O. 1952. Cited in Leplat (1966).

Davis, P.R.; Strubbs, D.A. Safe levels of manual forces for young males. *Applied Ergonomics* 1977, **8**, pp. 141, 219, **9**, p. 33.

De Felice, S. *Recherches sur l'Anthropologie des Francaises.* Masson, Paris, 1958.

De Greene, K.B. *Systems Psychology* McGraw-Hill, New York, 1970.

De Montmollin, M. Man-machine systems. *PUF Collection* Paris, 1967.

Desroches, M.; Cherel, G. Gas-tight cell and magnetic remote-controlled manipulator. *Proceedings of 9th HLE* 1961, p. 87.

Devjanin, E.A. *et al.* The six-legged walking robot capable of terrain adaptation. *Proceedings of 4th RO.MAN.SY* Warsaw, 1981, p. 44.

Devresse, N.; Da Costa, L.; De Buck, F. TELEM, vehicle for remote inspection and intervention in environments unhealthy or non-accessible for humans. *Proceedings of International Conference on Robotics and Remote Handling in the Nuclear Industry* Toronto, 1984.

Diaz, R.G.; Boykin, W.H. Teleoperator systems for nuclear applications in the 1980s. A technology assessment. *Proceedings of 27th RSTD Conference* 1979, p. 124.

Dombre, E.; Liégeois, A. Analysis of industrial robots. Relationships between geometry, performance and function. *IRIA/SURF* Report No. 79102, Montpellier, July, 1979.

Dupeyrat, B. Thesis, University of Paris VI, 1981.

Dupeyrat, B. *et al. Bulletin INRIA* No. 74, December, 1981.

Edahiro, K. Development of an underwater robot cleaner for marine live growth in power stations. *Proceedings of 1983 ICAR* Tokyo, September, 1983, p. 99.

Espiau, B. Detection, estimation de l'orientation et saisie d'une cible mobile par proximetrie optique. *IRISA Internal Publications* No. 143, January, 1981.

Espiau, B. Thesis, University of Rennes, June, 1982.

Espiau, B.; André, G. Using proximity sensors in computer-enhanced telemanipulation. *Proceedings of 4th RO.MAN.SY* Warsaw, 1981, p. 231.

Espiau, B.; André, G. Sensory-based control for robots and teleoperators. *Proceedings of 5th RO.MAN.SY* Udine, 1984, Kogan Page, London, 1985.

Espiau, B.; Catros, J.Y. Use of optical reflectance sensors in robotics applications. *IEEE SMC* Volume 10, No. 12, December, 1980, p. 903.

Espiau, B. *et al.* The Spartacus telethesis. *Proceedings of International Conference on Telemanipulators for the Physically Handicapped* IRIA, Rocquencourt, September, 1978.

Espiau, B.; André, G.; Borelly, J.J. Utilization of proximity data in teleoperation. *Proceedings of 1st ARA Journal* Poitiers, September, 1982, p. 32. *Proceedings of 2nd ARA Journal* Besancon, November, 1983.

Estabrook, N.B. Advanced search and work systems. *International Conference on Underwater Penetration* Edition Technip, Paris, December, 1982, p. 336.

Faugeras, P. *et al.* Carmen chemistry facility for reprocessing irradiated fuels in gas-tight shielded enclosures. *Proceedings of 11th HLE ANA Conference* New York, November, 1963, p.89.

Faverge, J.M. Work analysis and structure. *B.CERP* Volume 1, 1952, p. 1.

Feldmann, MJ.; White, J.R. Remotex, a new concept for efficient remote operations and maintenance in nuclear fuel reprocessing. *Proceedings of 28th RSTD Conference* November, 1980, p. 3.

Feldmann, M.J.; Hamel, W.R. The advancement of remote systems technology; past perspectives and future plans. *Proceedings of International Conference on Robotics and Remote Handling in the Nuclear Industry* Toronto, September, 1984I, p. 173.

Feldmann, M.J.; Hamel, W.R. The advancement of remote technology: past perspectives and the future. *Seminar on Remote Handling Equipment for Nuclear Fuel Cycle Facilities, IAEA* No. SR103, IAEA-OECD Nuclear Energy Agency, Oxford, 1984II.

Ferrell, W.R. Remote manipulator with transmission delay. *IEEE Transactions on Human Factors in Electronics, HFE* Volume 6, 1965, p. 24.

Ferrell, W.R. Delayed force feedback. *IEEE Transactions on Human Factors in Electronics, HFE* Volume 8, 1966, p. 449.

Ferrell, W.R.; Sheridan, T.B. Supervisory control of remote manipulation. *IEEE Spectrum* Volume 4, October, 1967, NASA Report SP144, 1967, p. 315.

Fiori, N.; Acquaviva-Lescoat, D. Contribution to the study of the man-manipulator interface at a simple manipulation task. *Proceedings of 31st RSTD Conference* Volume 1, June, 1983, p. 26.

Fiori-Savary, N.; Richardson, J. *Journal of Physiology* 1979, **75**, p. 43A.

Fiori, N.; Acquaviva-Lescoat, D.; Richardson, J. Mesure de la performance et evaluation de la charge de travail en course de la telemanipulation. *Proceedings of 2nd ARA Journal* Besancon, 1983.

Fitts, P.M. The information capacity of the human motor system in controlling the amplitude of movement. *J. Exp. Psychol.* 1954, **47**, p. 381.

Fitts, P.M. The information capacity of the human motor system in controlling the amplitude of movement. *Journal of Experimental Psychology* 1954, **47**, p. 391.

Flatau, C.R. Development of servo manipulators for high energy accelerator requirements. *Proceedings of 13th RSTD Conference* 1965, p. 29.

Flatau, C.R. Compact servo master-slave manipulator with optimized communication links. *Proceedings of 17th RSTD Conference* 1969, p. 154.

Flatau, C.R. Can mankind benefit from extension systems? *Proceedings of 18th RSTD Conference* 1970, p. 204.

Flatau, C.R. The manipulator as a means of extending our dextrous capabilities to larger and smaller scales. *Proceedings of 21st RSTD Conference* 1973, p. 47.

Flatau, C.R. SM229, a new compact servo master-slave manipulator. *Proceedings of 25th RSTD Conference* 1977, p. 169.

Flatau, C.R.; Da Costa, L. Remote vehicle for hostile environments. *Proceedings of 32nd RSTD Conference* 1984, Volume 2, p. 11.

Flatau, C.R.; Greeb, F.J.; Booker, R.A. Some preliminary correlations between control modes of manipulator systems and their performance indices. *Proceedings of 1st RMS Conference* Pasadena, 1972, p. 189.

Foote, A.L.; Dompe, R.S. Applications for head-aimed television systems. *Proceedings of 26th RSTD Conference* 1978, p. 83.

Foote, A.L. *et al.* A head-coupled television for remotely manned driving and manipulation tasks. *Proceedings of Symposium on Visually Coupled Systems: Developments and Applications* Aerospace Medical Division, Brooks Airforce Base, Texas, 1972.

Fournier, A. *et al.* Use of artificial visual feedback signals to control a remote manipulator. *Proceedings of ARA Journal* Toulouse, 1984, p. P1.

Fournier, R. Thesis, USTL, Montpellier, 1983.

Fraize, G. *Rapport Interne ARA, Pole Mecanique et Technologie* 1983.

Fraize, G.; Vertut, J. Coverage optimization of articulated manipulator within force and inertia acceptable limits. *Proceedings of 5th RO.MAN.SY* Udine, 1984, Kogan Page, London, 1985.

Frank, A.A. *Automatic Control System for Legged Locomotion Machine.* PhD Thesis, University of Southern California, May, 1968.

Frank, A.A. *et al.* Design of a manipulator arm for an interactive computer-controlled legged locomotion system. *Proceedings of Milwaukee Symposium on Automatic Computation and Control* USA, April, 1976.

Freedy, A.; Weltam, G. Performance improvement in remote manipulation with time delay by means of a learning system. *Proceedings of 1st RMS Conference* 1972.

Freedy, A. *et al.* A computer-based learning system for remote manipulator control. *IEEE, SMC* October, 1971.

Fuggit, B.; Uhrich, R.W. Underwater stereoscopic television and display realism. *NOSC.* TP358, 1976.

Fuller, G.M.; Zahn, H.S. The impact of the fusion reactor design approach on maintenance of first wall: blanket modules. *Proceedings of 26th RSTD Conference* 1978, p. 325.

Gaillard, J.P. The persistence of a visual dominance effect in telemanipulation task: a comparison between visual and electrotactile feedback. *Proceedings of 17th Annual Conference on Manual Control* 1981, p. 127.

Gaillard, J.P. La visualisation des efforts. *Proceedings of 2nd ARA Journal* Besancon, November, 1983.

Gaillard, J.P.; Quetin, N. Evaluation of sensory feedback information. *Proceedings of 1st ARA Journal* Poitiers, September, 1982, p. 81.

Galbiati, L. *et al.* A compact and flexible servo system for master-slave electric manipulator. *Proceedings of 12th HLE* 1964, p. 73.

Gardner, J.E. *et al.* Design and testing of a digitally controlled hydraulic actuation system for a walking vehicle mechanism. *Proceedings of 8th Applied Mechanism Conference* St Louis, September, 1983.

Germond, P. Building of a facility for the handling of kilocurie amounts of gamma emitters. *Proceedings of 7th HLE* 1959, p. 71.

Gibson, J.J. *The Senses Considered as Perceptual Systems* Houghton Mifflin, Boston, 1966.

Gilbreth, F.B. *et al. Management and Administration* Volume 8, 1924, p. 152.

Goertz, R.C. Manipulator systems development at ANL. *Proceedings of 12th RSTD Conference* 1964, p. 117.

Goertz, R.C. *et al.* The ANL model 3 master-slave manipulator design and use in a cave. *Proceedings of 9th HLE* 1961I, p. 121.

Goertz, R.C. *et al.* A technique for making extremely flexible electric cables. *Proceedings of 9th HLE* 1961II, p. 161.

Goertz, R.C. *et al.* On experimental head-controlled television system to provide viewing for a manipulator operator *Proceedings of 13th RSTD Conference* 1965, p. 57.

Goertz, R.C. *et al.* ANL Mark E 4A electric master-slave manipulator. *Proceedings of 14th RSTD Conference* 1966I, p. 115.

Goertz, R.C. *et al.* The ANL Mark TV2. An experimental five motion head-controlled television system. *Proceedings of 14th RSTD Conference* 1966II, p. 124.

Grandjean, E.; Vigliani, E. *Ergonomics Aspects of Visual Display Terminals* Taylor Francis, London, 1980.

Grenon, M. *Le Travail en Milieu Hostile* University Press of France, Paris, 1968.

Grisham, D.L.; Lambert, J.E. Why high capacity manipulators and lead glass window? *Proceedings of 29th RSTD Conference* Volume 1, 1981, p. 13.

Grisham, D.L.; Lambert, J.E. Human factor considerations in the monitor system. *ANS Transactions* Volume 44, June, 1983, p. 568. *Proceedings of 31st RSTD Conference* Volume 1, 1983.

Grisham, D.L. *et al.* Monitor 1979. *IEEE Transactions of Nuclear Science* Volume NS26-3, 1979I, p. 3376.

Grisham, D.L. *et al.* Monitor update 1979. *Proceedings of 27th RSTD Conference* 1979II, p. 268.

Grisham, D.L. *et al.* Monitor 1980: now they are two. *Proceedings of 28th RSTD Conference* Volume 2, 1980, p. 83.

Grunwald, A.J. Predictor symbology in computer-generated perspective display. *Proceedings of 17th Conference on Manual Control* June, 1981, p. 89.

Guinot, J.C.; Espiau, B. *1st Workshop NSF-ARA* NSF Int 8112603, Paris, 1982, p. 159.

Guinot, J.C.; Lallemand, J.P.; Zeghloul, S. Etude de la prehension et sens tactile. *Proceedings of 1st ARA Journal* Poitiers, 1982.

Guinot, J.C.; Lallemand, J.P.; Zeghloul, S. Analyse mecanique d'un prehenseur tridigital a capteur d'efforts, et commande. *Congres de la Societe Francaise de Mecanique* September, 1983I.

Guinot, J.C.; Lallemand, J.P.; Zeghloul, S. Etude d'un prehenseur tridigital avec sens tactile pour une operation d'assemblage. *Proceedings of 6th World Congress on the Theory of Machines and Mechanisms* New Delhi, December, 1983II, p. 1011.

Guittet, J. Thesis, Université de Droit, d'Economie et des Sciences d'Aix Marseille, 1985I.

Guittet, J. *La Robotique Medicale.* Hermes Publishing, Paris, 1985II.

Guittet, J.; Parent, M. The MAT.1 proposition for a microprocessor-based telemanipulator system. *Proceedings of 3rd RO.MAN.SY* Udine, 1978, p. 447.

Guittet, J. *et al.* Spartacus telethesis. *Proceedings of International Conference on Telemanipulators for the Physically Handicapped* IRIA, Rocquencourt, September, 1978.

Guittet, J.; Kwee, H.H.; Vertut, J. A computer-enhanced remote manipulator for severely handicapped operators. *Proceedings of 4th RO.MAN.SY* Warsaw, 1981, p. 256.

Gurfinkel, V.S.; Gurfinkel, E.V.; Schneider, A.Y. Walking robot with supervision control. *Proceedings of 3rd RO.MAN.SY* Udine, 1978, p. 576.

Haaken, L.W.; Olsen, R.A.; Jelatis, D.G. A gas-tight direct-coupled mechanical master-slave manipulator for alpha gamma facilities. *Proceedings of 10th HLE* 1962, p. 153.

Hamonet, C. *et al.* Orthesis and telemanipulators. *Proceedings of International Conference on Telemanipulators for Physically Handicapped* IRIA, Rocquencourt, September, 1978.

Hanafusa, H.; Asada, H. Stable prehension by a robot hand with elastic fingers. *Proceedings of 7th International Symposium on Industrial Robots* Tokyo, 1977, p. 361.

Hanafusa, H.; Asada, H. Adaptive control of a robot hand with elastic fingers for mechanical assembly. *Proceedings of 3rd RO.MAN.SY* Udine, 1978, p. 45.

Hanafusa, H. *et al.* Fine control of objects with articulated multifinger robot hand. *Proceedings of 1983 ICAR* 1983, p. 245.

Handlykken, M.; Turner, T. Control system analysis and synthesis for a six DOF universal force reflecting hand controller. *Proceedings of 19th IEEE Conference on Decision and Control* Albuquerque, 1980.

Heckman, P.J. Free swimming submersible test bed (EAVE WEST). *NOSC* Tr622, September, 1980.

Hensel, H. Thermoreceptors. *Annual Review of Pysiology* 1974, **36**.

Herndon, J.N.; Jelatis, D. The state of art model M2 maintenance system. *Proceedings of Robotics and Remote Handling in Hostile Environments, National Topical Meeting* ANS Gatlinburg, 1984.

Hill, J.W. Hydraulically powered lower limb orthosis. *Proceedings of 2nd RO.-MAN.SY* Warsaw, 1976I.

Hill, J.W. Comparison of seven performance measures in a time-delayed manipulation task. *IEEE Transactions on Systems, Man and Cybernetics* Volume SMC 6, No. 4, 1976II, p. 286.

Hill, J.W.; Salisbury, J.K. Study to design and develop remote manipulator systems. *SRI Annual Report* No. 2, 1977.

Hirose, S. A study of design and control of a quadruped walking vehicle. *Robotics Research* Special Issue, Volume 3, No. 2, 1984, p. 113.

Hirose, S.; Umetani, Y. Some considerations on a feasible walking mechanism as a terrain vehicle. *Proceedings of 3rd RO.MAN.SY* Udine, 1978, p. 357.

Hirose, S.; Umetani, Y. The basic motion regulation system for a quadruped walking machine. *Proceedings of ASME Conference* Det 34, 1980.

Homer, G.B. Mobile manipulator systems. *Proceedings of 14th RSTD Conference* 1966, p. 129.

Horne, R.A. Mantis, a compact mobile remote handling system for accelerator halls and tunnels. *Proceedings of 26th RSTD Conference* 1978, p. 55.

Horne, R.A.; Ekberg, E.L. Monitor, a versatile remote handling system. *Proceedings of 23rd RSTD Conference* 1975, p. 242.

Horne, R.A.; Ellefsplass, M. Long range, high speed remote handling at the CERN 26 GEV proton synchroton. *Proceedings of 23rd RSTD Conference* 1975, p. 247.

Horst, R.L. *et al.* Studies in three-dimensional viewing in teleoperated systems. *ANS Transactions* TANSAO 44, 1983, p. 567, *Proceedings of 31st RSTD Conference* Volume 1, 1983.

Hotta, I.; Ohtsuika, K.; Natori, K. Underwater bilateral servo manipulator, Japan Marine Science and Technology Center. *Proceedings of 7th International Symposium on Industrial Robots* Tokyo, 1977, p. 379.

Hugon, R. *CEA STEP* Internal Report, 1982.

Hugon, R. *CEA STEP* Internal Report, 1983.

Ichikawa, Y.; Ozaki, N.; Sadakome, K.I. Five-legged vehicle for remote systems in nuclear plants. *Proceedings of 30th RSTD Conference* Volume 2, 1982, p. 150.

Ishiro, Y. *et al.* Walking robot for underwater construction. *Proceedings of 1983 ICAR* Tokyo, September, 1983, p. 107.

Jacobsen, S.C. *et al.* An electro pneumatic actuation system for the Utah/MIT dextrous hand. *Proceedings of 5th RO.MAN.SY* Udine, 1984, Kogan Page, London, 1985.

Jelatis, D.G. Power-assisted 45 kg (100 lbs) grip system for master-slave manipulators. *Proceedings of 25th RSTD Conference* 1977, p. 158.

Jelatis, D.G.; Haacker, L.W.; Olsen, R.A. A rugged duty man-capacity master-slave manipulator. *Proceedings of 10th HLE* 1962, p. 153.

Jimenez, E. In-vessel manipulator for the Tokomak fusion test reactor. *Proceedings of 26th RSTD Conference* 1978, p.313.

Johannsen, G. Workload and work measurement. *Mental Workload* Plenum Press, New York, 1979, p. 3.

Johnsen, E.G. Telesensors, teleoperators and telecontrols for remote operations. *IEEE Transactions on Nuclear Science* Volume NS13, 1966, p. 14.

Johnsen, E.G. Advancements in teleoperator systems. *An AEC-NASA Technology Utilization Publication* NASA Report SP5081, 1970.

Johnsen, E.G.; Corliss, W.R. Teleoperators and human augmentation. *NASA Report* SP5047, 1967.

Johnsen, E.G.; Corliss, W.R. *Human Factors Application in Teleoperator Design and Operation.* Wiley, New York, 1971.

Jolly, D. *Proceedings of 1st ARA Meeting* Poitiers, 1982.

Kanedo, M. *et al.* Legged locomotion machine based on the consideration of degrees of freedom. *Proceedings of 5th RO.MAN.SY* Udine, 1984, Kogan Page, London, 1985.

Karhu, O.; Kansi, P.; Kuorinka, I. Correcting working postures in industry: a practical method for analysis. *Applied Ergonomics* 1977, 8-4, p. 199.

Kashihara, H. Development program of two arm bilateral servo manipulator system for nuclear fuel cycle facilities in PNC. *Proceedings of Robotics and Remote Handling in Hostile Environments, National Topical Meeting ANS* Gatlinburg, 1984, p. 245.

Kashihara, H. *et al.* Servo manipulator having new multijoint and ingenious wrist. *Proceedings of 1983 ICAR* Tokyo, September, 1983, p. 59.

Kato, I.; Tsuiki, M. Hydraulically powered biped walking machine with a high carrying capacity. *Proceedings of 4th International Symposium on External Control of Human Extremities* Dubrovnik, 1972.

Kato, I. *et al.* Information power machine with senses and limbs (robot I). *Proceedings of 1st RO.MAN.SY* Udine, Volume 1, 1973, p. 22.

Kato, I. *et al.* The realisation of the quasi-dynamic walking by the biped walking machine. *Proceedings of 4th RO.MAN.SY* Warsaw, 1981, p. 34.

Keegan, J.J. Alterations of the lumbar curve related to posture and seating. *Journal of Bone and Joint Surgery* 1953, **35A**, p. 580.

Keele, S.W. Movement control in skilled motor performance. *Psychology Bulletin* 1968, **70**, p. 387.

Kessis, J.J. *et al.* Walking robot multilevel architecture and implementation. *Proceedings of 4th RO.MAN.SY* Warsaw, 1981. p. 297.

Kessis, J.J. *et al.* Hexapod walking robots with artificial intelligence capabilities. *Proceedings of 5th RO.MAN.SY* Udine, 1984, Kogan Page, London, 1985.

Khatib, O,; Le Maître, J.F. Dynamic control of manipulators operating in a complex environment. *Proceedings of 3rd RO.MAN.SY* Udine, 1978, p. 267.

Kleinwächter, H. Syntelman. *Advancement in Teleoperator Systems* NASA Report SP5081, 1970, p. 138.

Kleinwächter, H.; Droge, K.M. The Syntelman teleoperator in nuclear plants. *Proceedings of 2nd Conference on Industrial Robot Technology* Volume 2, 1974, p. 13.

Kobayashi, A.; Nakamura, K. Rescue robot for fire hazard. *Proceedings of 1983 ICAR* Tokyo, September, 1983, p. 91.

Köhler, G.W. *Manipulator Type Book* Verlag Karl Thiemig, Munich, 1981.

Köhler, G.W.; Selig, M.; Salaska, M. Manipulator vehicles of the Nuclear Emergency Brigade in the Federal Republic of Germany. *Proceedings of 24th RSTD Conference* 1976, p. 196.

Konstantinov, M.S.; Genova, P.J. Workspace and manoeuvrability criteria for robots. *Proceedings of 4th RO.MAN.SY* Warsaw, 1981, p. 382.

Kosby, R.P. Remotely actuated television system. *Proceedings of 19th RSTD Conference* 1971, p. 172.

Kroczinski, P. The marine robot. *Proceedings of 14th International Symposium on Industrial Robots* Volume 1, Detroit, 1984, p. 9.

Kuban, D.P. An advanced remotely maintainable servo manipulator concept. *Proceedings of Robotics and Remote Handling in Hostile Environments, National Topical Meeting ANS* Gatlinburg, 1984, p. 407.

Kuban, D.P.; Perkins, G.S. Dual arm master controller concept. *Proceedings of Robotics and Remote Handling in Hostile Environments, National Topical Meeting ANS* Gatlinburg, 1984, p.433.

Kugath, D.A.; Wilt, D.R. Problems in selection of design parameters affecting manipulator performance. *Proceedings of 1st RO.MAN.SY* Udine, Volume 2, 1973, p. 169.

Kuspriyanto, P. Thesis, USTL, Montpellier, 1981.

Kwee, H.H.; Pannier, S. Medical robot for quadriplegics. *Proceedings of International Rehabilitation Workshop* Stanford University, August, 1981.

Kwee, H.H. Telethesis: assisted manipulation. *Proceedings of Workshop on Evaluation of Assistive Devices for Paralysed Persons* Council of European Communities, Milan, April, 1983.

Lambert, J.E.; Grisham, D.L. History of remote handling at LAMPF. *Proceedings of 30th RSTD Conference* Volume 1, 1982, p. 3.

Landau, L.; Lifchitz, E. Physique theorique. *Mecanique* Volume 1, MIR, Moscow, 1960.

Lefort, G.; Cazalis, J.P. A new type of air-tight tongs for master-slave manipulators. *Proceedings of 12th HLE* 1964, p. 167.

Lefort, G.; Vertut, J.; Cazalis, J.P. New interchangeable seals for ports in alpha beta gamma cells. *Proceedings of 11th HLE* 1963, p. 353.

Lefort, G. *et al.* A new type of chemistry cell for high specific activity at Fontenay-aux-Roses. *Proceedings of 12th RSTD Conference* 1964, p. 15.

Leifer, L. Rehabilitative robots. *Robotics Age* 1981, 3 (3), p. 95.

Leifer, L.; Sachs, J. Voice command of a six DOF manipulator. *Proceedings of Joint Automatic Control Conference* Denver, June, 1979, p. 1.

Leifer, L. *et al.* Robotics aids for the severely disabled: needs assessment. *Proceedings of International Conference on Telemanipulators for Physically Handicapped* IRIA, Rocquencourt, September, 1978.

Leister, P. DWK test facility for remote maintenance. First results of test program. *Proceedings of Robotics and Remote Handling Environments, National Topical Meeting ANS* Gatlinburg, 1984I, p. 227.

Leister, P. WA 350 German spent fuel reprocessing plant. Experiments in remote process equipment design. *Seminar on Remote Handling Equipment for Nuclear Fuel Cycle Facilities, IAEA* No. SR103, IAEA-OECD Nuclear Energy Agency, Oxford, 1984II.

Lepers, B. Introducing safety in the conception and design of a telemanipulator. Application to the design of a surgical micromanipulator. *Proceedings of System Engineering III* Coventry, September, 1982.

Lepers, B. Thesis, USTL, Lille, 1983I.

Lepers, B. Micromanipulateurs stereotaxiques dans la domaine chirurgical. *Journal of SEE* Rennes, November, 1983II.

Lepers, B.; Diaz, C. Conception of a surgical micromanipulator. *Proceedings of 3rd International Conference on Mechanics in Medicine and Biology* Compiegne, July, 1982.

Leplat, J. *B. Cerp.* 1966, 1, pp. 351-358.

Leplat, J. *PUF* 1970, 3, p. 74.

Leplat, J.; Cuny, X. *PUF Collection* Paris, 1977.

Leroi-Gourhan, A. *The Gesture and the Words*, 1966.

Lestienne, F. Thesis, USTL, Lille, 1974.

Lhote, F.; Kauffmann, J.; André, P.; Taillard, J. *Robot Components and Systems* Robot Technology, Volume 4, Kogan Page, London, 1984.

Liege, T. Thesis, USTL, Montpellier, 1980.

Liégeois, A. *Performance and Computer-Aided Design* Robot Technology, Volume 7, Kogan Page, London, 1985.

Liégeois, A.; Dombre, E. Analysis of industrial robots. Relationships between geometry, performance and function. *IRIA/SURF* Report No. 79102, July, 1979.

Loewenstein, W.R. *Handbook of Sensory Physiology* Volume 1, 1971.

Lopez Krahe, Thesis, University of Paris VIII, April, 1979.

Lowe, D.B. Applications of robot technology in the inspection and repair of active nuclear plants. *Proceedings of 29th RSTD Conference* Volume 2, 1981, p. 133.

Lukishow, G.I.; Milosardin, Y.V. Criteria for evaluation of kinematics characteristics of master-slave manipulator and a method for the study of their space performance. *Proceedings of 1st RO.MAN.SY* Udine, Volume 2, 1973, p. 191.

McGhee, R.B. Control of legged locomotion systems. *Proceedings of 18th Joint Automatic Control Conference* 1977, p. 205.

McGhee, R.B.; Orin, D.E. An interactive computer-control system for a quadruped robot. *Proceedings of 1st RO.MAN.SY* Udine, Volume 1, 1973, p. 25.

McGhee, R.B.; Orin, D.E. A mathematical approach to control joint positions and torques in legged locomotion systems. *Proceedings of 2nd RO.MAN.SY* Warsaw, 1976, p. 225.

McGhee, R.B.; Wandhi, J. Optimization of support state sequences for legged locomotion systems. *Technical Notes* No. 22, Communication and Control Systems Laboratory OSU, Columbus, March, 1978.

McGhee, R.B. *et al.* Real time computer control of a hexapod vehicle. *Proceedings of 3rd RO.MAN.SY* Udine, 1978, p. 323.

McGhee, R.B. *et al.* A hierarchically structured system for computer control of a hexapod walking machine. *Proceedings of 5th RO.MAN.SY* Udine, 1984, Kogan Page, London, 1985.

McGovern, D.E. Comparison of two manipulators using a standard task of varying difficulty. *ASME Publication* 1974, p. 17.

Maeda, M. *et al.* Remote maintenance equipment for hot cell facilities. *Proceedings of 30th RSTD Conference* Volume 2, 1982, p. 132.

Malvache, N. *Le Point sur l'Automatique Humaine* LAIH, University of Valenciennes, 1981.

Marchal, R.; Charles, J.; Vertut, J. *Recherches et Technologies en Milieux Extremes* CESTA, Paris, 1983, p. 41.

Martin, H.L. Control and electronic subsystems for the advanced servo manipulator. *Proceedings of Robotics and Remote Handling in Hostile Environments, National Topical Meeting ANS* Gatlinburg, 1984, p. 383.

Martin, H.L.; Satterlee, P.E.; Bolling, B.J. Distributed digital processing for servo manipulator control. *Proceedings of 30th RSTD Conference* Volume 2, 1982, p. 117.

Massion, J. Le système pyramidal: données récentes. *Du Contrôle moteur à l'organisation du Geste* Volume 1, H. Hecaen and M. Jeannerod, Eds, Masson, Paris, 1978, p. 31.

Matsubara, T. *et al.* Development of a mobile manipulator for maintenance and inspection in containment vessels of nuclear power plants. *Seminar on Remote Handling Equipment for Nuclear Fuel Cycle Facilities, IAEA* No. SR103, IAEA-OECD Nuclear Energy Agency, Oxford, 1984.

Mechanix Illustrated September, 1964, p. 65.

Micaelli, A. Thesis, University of Paris XI, Orsay, June, 1982.

Michel, J.L.; Le Roux, H. Epaulard: deep bottom survey now with acoustic remote-controlled vehicle. *Ocean 1981 Conference Record, The Ocean: An International Workplace* Volume 1, Boston, September, 1983, p. 99.

Mischke, J.; Hendrich, K. Remote maintenance concept for the DWK spent fuel reprocessing plant. *Proceedings of Robotics and Remote Handling in Hostile Environments, National Topical Meeting ANS* Gatlinburg, 1984, p. 219.

Mischke, J.; Leister, P. DWK test facility for remote maintenance. *Proceedings of Robotics and Remote Handling in Hostile Environments, National Topical Meeting ANS* Gatlinburg, 1984, p. 227.

Mocci, V.; Petternella, M.I.; Salinari, S. *Experiments with Fixed Gait* Institute of Automation, University of Rome, Report No. 2.12, 1972.

Monod, H. *Precis de Physiologie du Travail* 2nd Edition, J. Scherrer *et al.* Eds, Masson, Paris, 1981I, p. 107.

Monod, H. *Precis de Physiologie du Travail* 2nd Edition, J. Scherrer *et al.* Eds, Masson, Paris, 1981II, p. 159.

Moreau, J.C. *Utilization de la Commande Vocale pour Commande de l'Environnement* INRIA Bulletin No. 74, 1981.

Morecki, A.; Eckiel, J. Concept of anthropomorphism in description of androidal robots. *Proceedings of 3rd RO.MAN.SY* Udine, 1978, p. 98.

Morecki, A.; Bianchi, G.; Kedzior, K. (Eds) *Proceedings of 5th RO.MAN.SY* Udine, 1984, Kogan Page, London, 1985.

Mosher, R.S. An electrohydraulic bilateral servo manipulator. *Proceedings of 8th HLE* 1960, p. 252.

Mosher, R.S. *Handyman to Hardiman* Society of Automotive Engineers, SAE Report 680088, 1967.

Mosher, R.S. *Exploring the Potential of a Quadruped* Society of Automotive Engineers, SAE Report 690191, 1969.

Mosher, R.S.; Wendel, B. Force reflecting electro-hydraulic servo manipulator. *Electro Technology* 1960, **66**, p. 138.

Muller, D.P. *An Evaluation of Resolved Motion Rate Control for Remote Manipulators* MS Thesis, MIT, 1973.

Murphy, K.N.; Raibert, M.H. Trotting and bounding in a planar two-legged model. *Proceedings of 5th RO.MAN.SY* Udine, 1984, Kogan Page, London, 1985.

Nakano, E.; Koyachi, N. An advanced mechanism of the omnidirectional vehicle (ODV) and its application to the working of a wheelchair for the disabled. *Proceedings of 1983 ICAR* Tokyo, September, 1983, p. 277.

Nakano, E. *et al.* Cooperational control of the anthropomorphous manipulator 'Melarm'. *Proceedings of 4th International Symposium on Industrial Robots* November 1974, p. 251.

National Geographic Magazine 1983, **164**, No. 1, p. 104.

Nevins, J.L. *Teleoperator Technology: Past, Present and Future* MIT Draper Laboratory Report E26240, 1972.

Nodland, W.E.; Eward, T.E.; Bendiner, W.P. SPURV II: An unmanned, free swimming, submersible developed for oceanographic research. *Ocean 1981 Conference Record: The Ocean: An International Workplace* Volume 1, Boston, September, 1981, p. 92.

NOSC *Ocean Technology: A Digest* Technical Document No. 149, San Diego, 1978.

Nuclear System Associates *ORNL: Hot Experimental Facility Remote Maintenance Time and Study* Report No. N78018R 3, Brea, 1978.

Ogle, K.N. *Researches in Binocular Vision* Hofner, New York, 1950.

Ogo, K.; Ganse, A.; Kato, I. Quasi dynamic walking of biped aiming at completion of steady walking. *Proceedings of 3rd RO.MAN.SY* Udine, 1980, p. 340.

Okada, T. Computer control of multi-finger systems. *Proceedings of 6th International Symposium on Industrial Robots* Tokyo, 1979.

Okoshi, T. *Three-Dimensional Imaging Techniques* Academic Press, New York, 1976.

Ombredane, A.; Faverge, J.M. *L'analyse du Travail: Facteurs d'Economie Humaine et de Productivite* PUF, Paris, 1955.

Orin, D.E.; McGhee, R.B. Dynamic computer simulation of robotic mechanism. *Proceedings of 4th RO.MAN.SY* Warsaw, 1981, p. 186.

Orin, D.E.; Oh, Y. A mathematical approach to the problem of force distribution in locomotion and manipulator systems closed kinematic chains. *Proceedings of 3rd RO.MAN.SY* Udine, 1978, p. 1.

Orin, D.E.; McGhee, R.B.; Jaswa, V.C. Interactive computer control of a six-legged robot vehicle, terrain adaptability and energy. *Proceedings of IEEE Conference on Decision and Control* December, 1976.

Ozaki, N. Remotely operated manipulator vehicle for nuclear facilities. *Proceedings of Robotics and Remote Handling in Hostile Environments, National Topical Meeting ANS* Gatlinburg, 1984, p. 123.

Ozaki, N. *et al.* Robotics approach to remote inspection and maintenance in nuclear power plants. *Proceedings of 30th RSTD Conference* Volume 1, 1982, p. 35.

Ozaki, N. *et al.* Teleoperated mobile robot for remote maintenance in nuclear facilities. *Proceedings of 1983 ICAR* Tokyo, September, 1983, p. 67.

Ozaki, S.; Ptoh, K.; Inagaki, S. Standardization and characterization relating to industrial robots. *Proceedings of 7th International Symposium on Industrial Robots* Tokyo, 1977, p. 452.

Ozguner, F.; Tsai, S.J.; McGhee, R.B. Rough terrain locomotion by an hexapod robot using binocular ranging system. *Proceedings of 1st International Symposium on Robotics Research* Bretton Woods, September, 1983, MIT Press, Boston, 1984.

Ozguner, F.; Tsai, S.J.; McGhee, R.B. An approach to the use of terrain: preview information in rough terrain locomotion by an hexapod walking machine. *Robotics Research* Special Issue, Volume 3, No. 2, 1984, p. 134.

Pannier, S. Electronic help for the handicapped. *Proceedings of International Rehabilitation Workshop* Stanford University, August, 1981.

Paillard, J. *Du Schema Corporel a l'Espace Ecologique* PUF, Paris, 1972.

Parent, M.; Laurgeau, C. *Logic and Programming* Robot Technology, Volume 5, Kogan Page, London, 1984.

Paslack, V.; Rösler, H. Medical manipulators. General considerations on the design criteria of manipulative technical aid for quadriplegics. *Proceedings of 1st RO.-MAN.SY* Volume 1, Udine, 1973, p. 159.

Paslack, V.; Rösler, H. Design and control of a manipulator for tetraplegics. *Proceedings of 2nd RMS Conference* 1975.

Pech, J.; Peruchon, E. *Performances de l'Organe Terminal d'un Manipulateur* IRIA Report No. 76019, July, 1977.

Pepper, R.L., 1984, see Spain, E.H.; Pepper, R.L., 1984.

Peruchon, E. *Contribution to Carrying Out Artificial Tactile Functions Based on Analysis of the Human System* Thesis, Montpellier, 1979.

Peruchon, E.; Micallef, J.P. *Proceedings of 1st ARA Journal* Poitiers, September, 1982, p. 72.

Peruchon, E.; Micallef, J.P. An evaluation method for manipulator systems. *Proceedings of 2nd ARA Journal* Toulouse, 1984, p. H1.

Pesanti, P.; Cherel, G. French master slaves. *Proceedings of 7th HLE* 1959, p. 177.

Pierrel, J.M. *Contribution a la Comprehension Automatique du Discours Continu* Thesis, University of Nancy, 1975.

PMDS *Glossaire des Structures et Equipments des Protection pour les Installation Nucleaires* The CEA, Saclay, 1982.

Pott, G.; Stocksclader, F. Basic plan and special box techniques for the alpha-beta-gamma hot laboratory within the THTH project. *Proceedings of 14th RSTD Conference* 1966, p. 11.

Potts, C.W. Transistorized servo system for master slave electric manipulators. *Proceedings of ith HLE* 1961, p. 154.

Pradin, B. Thesis, UPS, Toulouse, 1971.

Quetin, N. *Remote Manipulation Experience of Two Tetraplegics.* Spartacus Project, IRIA Report No. HL021, 1978.

Quetin, N.; Gaillard, J.P.; Barbier, R. Evaluation of sensory feedback signals. *Proceedings of 1st ARA Journal* Poitiers, June, 1982.

Rabischong, P. *et al.* The OMAMI project perspectives. *Actualites en Reeducation Fonctionnelle et Readaptation* Masson, Paris, 1978, p. 179.

Raibert, M.H. Dynamic stability and resonance in a one-legged hopping machine. *Proceedings of 4th RO.MAN.SY* Warsaw, 1981, p. 352.

Raibert, M.H.; Sutherland, I.E. Machines that walk. *Scientific American* 1983, **248**, p. 44.

Raibert, M.H. *et al.* Dynamically stable legged locomotion. *1st Annual Report CMURI* January, 1982.

Raibert, M.H. *et al.* Experiments in balance with 3D one-legged hopping machine. *Robotics Research* Special Issue, Volume 3, No. 2, 1984, p. 75.

Ranadive, V.; Sheridan, T.B. Video framerate, resolution and grayscale trade-offs for undersea telemanipulator control. *Proceedings of 17th Conference on Manual Control* UCLA, 1981, p. 77.

Raynal, A.P.; Closson, Y.J.L. EDF graphite-gas Chinon A3, operation ISIS. *Seminar on Remote Handling Equipment for Nuclear Fuel Cycle Facilities, IAEA* No. SR103, IAEA-OECD Nuclear Energy Agency, Oxford 1984.

Rebiffe, R.; Zayana, O.; Tarriere, C.L. Report DRME. *Lab. Physiol. Biomécaniq.* 1966, **1**.

Renaud, M. *Mechanism and Machine Theory* 1980, **15**, p. 81.

Roger, D.; Millot, P. Ergonomie et robotique: le problème du dialogue homme-robot. *Carrefour sur la robotique Industrielle* INSA, Lyon, 1980.

Rösler, H.; Paslack, V. Manipulators for tetraplegics in adapted environment. *Proceedings of 2nd RO.MAN.SY* Warsaw, 1976, p. 234.

Roth, B. Performance evaluation of manipulation from a kinematic point of view. *Cours de Robotique* Volume 1, IRIA, 1976, p. 235.

Rovetta, A. On the prehension of the human hand. *MMT* Volume 14, No. 6, 1979, p. 385.

Sager, P.H. *et al. Proceedings of Fusion Engineering Device: Remote Maintenance Equipment* Workshop ORNL/TA 7769, 1981.

Sager, P.H. *et al. Fusion Engineering Device: Baseline Engineering Studies Report* ORNL/FEC/82/2, 1982.

Salisbury, J.K. *Kinematic and Force Analysis of Articulated Hand* PhD Thesis, Stanford, May, 1982.

Salisbury, J.K. Interpreting contact geometries from force measurements. *Proceedings of 1st International Symposium on Robotics Research* Bretton Woods, September, 1983, MIT Press, Boston, 1984.

Salisbury, J.K.; Craigh, J.J. Articulated hands: force control and kinematic issues. *International Journal of Robot Research* 1982, **2**, No. 2, p. 4.

Salisbury, J.K.; Roth, B. Kinematic and force analysis of articulated hands. *Journal of Mechanism Transmission and Automation Design* 1983, **105**, p. 35

Satterlee, P.E. Control software architecture and operating modes of the model M2 maintenance system. *Proceedings of Robotics and Remote Handling in Hostile Environments, National Topical Meeting ANS* Gatlinburg, 1984, p. 355.

Schmidt, R.A. A scheme theory of discrete motor skills learning. *Psychological Review* 1975, **82**, p. 225.

Schmidt, R.A. *et al.* Motor output variability: a theory for the accuracy of rapid motor acts. *Psychological Review* 1979, **86**, p. 415.

Schneider, A.Y. The six-legged walking robot with supervisory control. *Studies of Robotic Studies, HAYKA* Moscow, 1982.

Seamone, W. *et al.* Microprocessor controlled robotic arm workable system with wheelchair chin controller compatibility. *Proceedings of International Conference on Telemanipulators for Physically Handicapped* IRIA, Rocquencourt, September, 1978.

Sheridan, T.B. *Workshop on Performance Evaluation of Programmable Robots and Manipulators* NBS Special Publication 459, 1975.

Sheridan, T.B. Preview of models of the human monitor/supervisor. *Monitoring Behaviour and Supervisory Control* Plenum Press, New York, 1976, p. 175.

Sheridan, T.B. *Supervisory Control of Remote Manipulators, Vehicles and Dynamic Processes Experiments on Command and Display Aiding* Man-Machine System Laboratory, MIT, March, 1983.

Sheridan, T.B.; Ferrell, W.R. Remote manipulative control with transmission delay. *Proceedings of IEEE Transactions on Human Factors and Electronics* Volume 4 HFE, September, 1963, p. 25.

Sheridan, T.B.; Ferrell, W.R. *Man-Machine System* MIT Press, Boston, 1974.

Sheridan, T.B.; Verplank, W.L. *Human and Computer Control of Undersea Teleoperators* Man-Machine System Laboratory, MIT, July, 1978.

Shigley, J.E. *The Mechanics of Walking Vehicles* Land Locomotion Laboratory Report No. LL71, US Army Tank Automotive Command, Warren, 1960.

Shimano, B.; Roth, B. Ranges of notions of manipulators. *Proceedings of 2nd RO.MAN.SY* Warsaw, 1976, p. 18.

Shinohara, Y. Experimental remote handling system. *Proceedings of Robotics and Remote Handling in Hostile Environments, National Topical Meeting ANS* Gatlinburg, 1984, p. 117.

Simon, J. *et al.* Design of the Fermilab Neutrino Remote Target Maintenance System. *Proceedings of 23rd RSTD Conference* 1975, p. 32.

Song, S.M.. Waldron, K.J.; Kinzel, G.L. Computer-aided design of legs for a walking machine. *Proceedings of 8th Applied Mechanism Conference* St Louis, September, 1983.

Spain, E.H.; Pepper, R.L. Visual performance and fatigue with stereoscopic television displays. *Proceedings of Robotics and Remote Handling in Hostile Environments, National Topical Meeting ANS* Gatlinburg, 1984, p. 61.

Spampinato Remote maintenance equipment for the Fusion Engineering Device (FED). *Proceedings of 30th RSTD Conference* Volume 1, 1982, p. 64.

Spampinato *et al.* Three maintenance equipment concepts for the Fusion Engineering Device (FED). *Proceedings of 30th RSTD Conference* Volume 2, 1982, p. 184.

Stang, L.G. Rectilinear manipulator BNL model 4. *Proceedings of 7th HLE* April, 1959, p. 169.

Starr, G.P. *A Comparison of Master Slave and Rate Control in Presence of a Transmission Time Delay* PhD Thesis, Stanford, 1978.

Starr, G.P. Supervisory control of remote manipulation: a preliminary evaluation. *Proceedings of 17th Annual Conference on Manual Control* 1981, p. 95.

Stépourjine, R. A hydraulic modular robot to help seriously motor handicapped people. *Proceedings of 5th World Congress on the Theory of Machines and Mechanism* Volume 2, July, 1979, p. 1527.

Stojilkovic, Z.; Clot, J. Integrated behaviour of artificial skin. *IEEE Transactions* Volume EME-24, No. 4, July, 1977.

Stoughton, R.S.; Martin, H.L. Automatic camera tracking for remote manipulators. *Proceedings of Robotics and Remote Handling in Hostile Environments, National Topical Meeting ANS* Gatlinburg, 1984, p. 383.

Sutherland, I.E.; Ullner, M.K. Footprint in the aspalt. *Robotics Research* Special Issue, Volume 3, No. 2, 1984, p. 29.

Suzuki, M.; Ozaki, N.; Ichikawa, Y. A bilateral servo manipulator for remote maintenance in nuclear facilities. *Proceedings of 30th RSTD Conference* Volume 2, 1982, p. 138.

Tachi, S. *et al.* Study on tele existence (I) design and evaluation of a visual display with sensation of presence. *Proceedings of 5th RO.MAN.SY* Udine, 1984, Kogan Page, London, 1985.

Takanishi, A.; Naito, G.; Ishida, M.; Kato, I. Realization of plane walking by the biped walking robot WL-10R. *Proceedings of 5th RO.MAN.SY* Udine, 1984, Kogan Page, London, 1985, p. 383.

Takanori, O. *Three-Dimensional Imaging Techniques* Academic Press, New York, 1976.

Talkington, H. *Manned and Remotely Submersible Systems: A Comparison* Technical Paper No. 511, Naval Undersea Center, NUC, 1976.

Tani, K. *Supervisory Control of Remote Manipulation with Compensation for Moving Target* MIT Man-Machine Systems Laboratory Report, 1980.

Taylor, C.L. The biomechanics of the normal and of the amputated upper extremity. *Human Limbs and Their Substitutes* Klopsteg and Wilson, Eds, McGraw Hill, New York, 1954.

Taylor, H.J. A large scale manipulator for space shuttle payload handling – The Remote Manipulator System (RMS). *Proceedings of 26th RSTD Conference* ANS, 1978, p. 90.

Tesar, D.; Lipkin, H. *Assessment for the Man-machine Interface between the Human Operator and the Robotic Manipulator* Internal Report, Center of Intelligent Machines and Robotics, University of Florida, 1978I.

Tesar, D.; Lipkin, H. *Assessment for the Physical Structure and Hardware of General Robotics Manipulator Systems* Internal Report, Center of Intelligent Machines and Robotics, University of Florida, 1978II.

Thring, M.W. Telechiric mining. *Proceedings of 2nd RO.MAN.SY* Warsaw, 1976, p. 459.

Tomizawa, F.; Ozaki, N.; Sato, C. Infrared wireless telecommunication system for a remotely controlled robot in nuclear plans. *Proceedings of 31st RSTD Conference* Volume 1, 1983, p. 82.

Turner, 1927, cited in Monod, 1981II.

Uematsu, K. Remote system technology developments related to nuclear fuel cycle facilities in Japan. *Seminar on Remote Handling Equipment for Nuclear Fuel Cycle Facilities IAEA* No. SR103, IAEA-OECD Nuclear Energy Agency, Oxford, 1984.

Uricht, R.W. *Manipulator Development at the NUC* Technical Paper No. 553, Naval Undersea Center (now NOSC), 1977.

Vertut, J. New types of heavy manipulators. *Proceedings of 10th HLE* 1962, p. 185.

Vertut, J. Through the wall master-slave manipulator with indexing for forward and backward movement. *Proceedings of 12th RSTD Conference,* 1964, p. 67.

Vertut, J. European advances in teleoperators. *Proceedings of Advances in Teleoperator Systems* An AEC-NASA Publication, NASA SP5081, 1970, p. 209.

Vertut, J. Intervention vehicle Virgule and electronic master slave telemanipulator MA22. *Proceedings of 2nd Conference on Industrial Robot Technology* 1973I, p. C3-21.

Vertut, J. Contribution to analyse manipulator morphology, coverage and dexterity. *Proceedings of 1st RO.MAN.SY* Udine, Volume 1, 1973II, p. 277.

Vertut, J. Analytical zoning of manipulator coverage. *Proceedings of 21st RSTD Conference* 1973III, p. 28.

Vertut, J. Contribution to define a dexterity factor for manipulator. *Proceedings of 21st RSTD Conference* 1973IV, p. 38.

Vertut, J. Experience and remarks on manipulator evaluation. *Workshop on Performance Evaluation of Programmable Robots and Manipulators* NBS Special Publication 459, 1975, p. 97.

Vertut, J. *Remarks on Manipulator Classification, and Reversibility, Backlash, Inertia Balance and Friction Trade-off* CEA-STEP Internal Report, 1976.

Vertut, J. Control of master-slave manipulators and force feedback. *Proceedings of 1977 Joint Automatic Control Conference IEEE* San Francisco, 1977, p. 172.

Vertut, J. Conception of the MAT.1 manipulator. Spartacus, project, IRIA, Rocquencourt, September, 1978.

Vertut, J. MA23M contained servo manipulator with television camera, PICA and PIADE telescopic supports, with computer-integrated control. *Proceedings of 28th RSTD Conference* Volume 2, 1980, p. 13.

Vertut, J. Advances in computer-aided teleoperation systems (CATS) in the frame of the French ARA project. *Proceedings of 1983 ICAR* Suppl., Tokyo, September, 1983, p. 13.

Vertut, J.; Coiffet, P. Bilateral servo manipulator MA23 in direct mode and via optimized computer control. *Proceedings of 2nd RMS Conference* Volume 12, No. 5, 1977.

Vertut, J.; Flatau, C.R. MA22 compact bilateral servo master manipulator. *Proceedings of 20th RSTD Conference* 1972, p. 296.

Vertut, J.; Guilbaud, J.P. Virgule: variable geometry wheeled teleoperator. *Proceedings of 20th RSTD Conference* 1972, p. 303.

Vertut, J.; Liégeois, A. General design criteria for manipulator. *MMT* Volume 16, No. 1, 1981, p. 65.

Vertut, J.; Lefort, G.; Cazalis, J.P. Master-slave manipulator, model through the ceiling with index motion and alpha beta gamma shielding. *Proceedings of 11th RSTD Conference* 1963, p. 191.

Vertut, J.; Charles, J.; Coiffet, P.; Petit, M. Advances of the new MA23 force reflecting manipulator system. *Proceedings of 2nd RO.MAN.SY* Warsaw, 1976I, p. 307.

Vertut, J. *et al.* The MA23 bilateral servo manipulator systems. *Proceedings of 29th RSTD Conference* 1976II, p. 175.

Vertut, J. *et al.* Vehicles with wheels and legs, the in-pipe remote inspection vehicle and its family. *Proceedings of 3rd RO.MAN.SY Conference* 1978I, p. 476.

Vertut, J.; Marchal, P.; Moreau, J.P. Vehicules d'intervention pour grande immersion. Vehicule PLA 2. *Comptes Rendus Table Ronde Vecteurs Sous Marins* ASTEO, Paris, October, 1978II.

Vertut, J. *et al.* Protection transfer and maintenance of the MA23 bilateral servo manipulator. *Proceedings of 26th RSTD Conference* Volume 2, 1978III, p. 125.

Vertut, J.; Micaelli, A.; Marchal, P.; Guittet, J. Short transmission delay in a force reflective bilateral manipulator. *Proceedings of 4th RO.MAN.SY* Warsaw, 1981, p. 269.

Vertut, J.; Hugon, R.; Coiffet, P. Human factors aspects of computer-enhanced teleoperation. *Proceedings of 31st RSTD Conference* Volume 1, 1983I, p. 40.

Vertut, J. *et al.* Coordination camera telemanipulator by microprocessor. *Proceedings of 6th World Congress on the Theory of Machines and Mechanism* New Delhi, 1983II, p. 1071.

Vertut, J. *et al.* Advances in computer-aided bilateral manipulator system. *Proceedings of Robotics and Remote Handling in Hostile Environments, National Topical Meeting ANS* Gatlinburg, 1984I, p. 367.

Vertut, J. *et al.* Sensor-aided and computer-aided bilateral teleoperator system. *Proceedings of 5th RO.MAN.SY* Udine, 1984II, Kogan Page, London, 1985.

Vertut, J. *et al.* The MA23M computer-aided maintenance servo manipulator. *Proceedings of 32nd RSTD Conference* 1984III.

Vertut, J.; Millies, Rouyer, J.L. Nuclear robotics. *Seminar on Remote Handling Equipment for Nuclear Fuel Cycle Facilities, IAEA* No. SR103, IAEA-OECD Nuclear Energy Agency, Oxford, 1984IV.

Vertut, J.; Rouyer, J.L.; Clement, G. The teletrans concept for remote maintenance in Tokomak reactors. *Proceedings of International Conference on Robotics and Remote Handling in the Nuclear Industry* Toronto, 1984V.

Vertut, J. *et al.* Computer-aided teleoperation. *Proceedings of ARA Journal* Toulouse, September, 1984VI, p. A1.

Vidal, P. Teleoperation dans l'industrie agro-alimentaire. *SEE Journal* Rennes, November, 1983.

Vidal, P. *et al.* Man and robot task sharing. *Proceedings of ARA Journal* Poitiers, 1982, p. 54.

Vidal, P. *et al.* Supervisory system and the allocation of dynamic tasks. *Proceedings of ARA Meeting* Besancon, 1983, p. E1.

Vidal, P. *et al.* Supervisory system and the dynamic allocation of tasks. *Proceedings of ARA Meeting* Toulouse, 1984, p. E1.

Villamosa, H.; Ardebili, T.; Chirouze, M. Transformations de coordonnees pour robots et telemanipulateurs. *Proceedings of 1st ARA Journal* Poitiers, September, 1982, p. 20.

Vohnout, V.J.; Alexander, K.S.; Kinzel, G.L. The structural design of the legs for a walking vehicle. *Proceedings of 8th Applied Mechanism Conference* St Louis, 1983.

Waldron, K.J.; Kinzel, G.L. The relationship between actuator geometry and mechanical efficiency in robots. *Proceedings of 4th RO.MAN.SY* Warsaw, 1981.

Waldron, K.J. *et al.* Configuration design of the adaptive suspension vehicle. *Robotics Research* Special Issue Volume 3, No. 2, 1984I, p. 37.

Waldron, K.J. *et al.* Mechanical and geometrical design of the adaptive suspension vehicle. *Proceedings of 5th RO.MAN.SY* Udine, 1984II, Kogan Page, London, 1985.

Warnecke, H.J.; Brodbeck, B. Test stand for industrial robots. *Proceedings of 7th International Symposium on Industrial Robots* Tokyo, 1977, p. 443.

Wernli, R.L. Deep ocean applications of manipulators and work systems. *Proceedings of 27th RSTD Conference* 1979, p. 110.

Whitney, D.E. Resolved Motion Rate Control of manipulators and prosthesis. *IEEE Transactions on Man-Machine Systems* MMS 10, June, 1969.

Wilson, K.B. Servoarm. A water hydraulic master-slave manipulator. *Proceedings of 23rd RSTD Conference* 1975, p. 233.

Wilt, D.R. *et al.* An evaluation of control modes in high gain manipulator systems. *Mechanism and Machine Theory* Volume 12, 1977, p. 373.

Witham, A.; Foot, A.L. Design and application of remote manipulator system. *Proceedings of 26th RSTD Conference* 1978, p. 76.

Witham, A.; White, P.; Garin, J. Rigid hoist articulated grapple system development for enhanced remote maintenance. *Proceedings of 27th RSTD Conference* 1979, p. 270.

Yahazaki, T. *et al.* Development philosophy for an overall function evaluation system for the advanced robots. *Proceedings of 1983 ICAR* Tokyo, 1983, p. 115.

Yamamoto, M.; Inada, E. Remote maintenance equipment for hot cell facilities. *Proceedings of 30th RSTD Conference* Volume 2, 1982, p. 132.

Zeghloul, S. Thesis, University of Poitiers, October, 1983.

Index